Mandy Haggith

THE LOST ELMS

A Love Letter to Our Vanished Trees –
and the Fight to Save Them

WILDFIRE

Copyright © 2025 Mandy Haggith

The right of Mandy Haggith to be identified as the Author of
the Work has been asserted by her in accordance with the
Copyright, Designs and Patents Act 1988.

First published in 2025 by Wildfire
An imprint of Headline Publishing Group Limited

This paperback edition published in 2026

Apart from any use permitted under UK copyright law, this publication may
only be reproduced, stored, or transmitted, in any form, or by any means,
with prior permission in writing of the publishers or, in the case of
reprographic production, in accordance with the terms of licences
issued by the Copyright Licensing Agency.

Cataloguing in Publication Data is available from the British Library.

Paperback ISBN 978 1 0354 1234 1

Designed and typeset by EM&EN

Headline Publishing Group Limited
An Hachette UK Company
Carmelite House
50 Victoria Embankment
London EC4Y 0DZ

The authorised representative in the EEA is Hachette Ireland,
8 Castlecourt Centre, Dublin 15, D15 XTP3, Ireland (email: info@hbgi.ie)

www.headline.co.uk
www.hachette.co.uk

Mandy Haggith lives in a remnant of ancient rainforest in northwest Scotland and spent 20 years as a forest activist, from award-winning local campaigns in Scotland all the way to the United Nations. She is now an honorary research fellow and lecturer in creative writing at the University of the Highlands and Islands. She is the author of *Paper Trails: From Trees to Trash – The True Cost of Paper*, five novels and six poetry collections, and editor of the tree poem anthology, *Into the Forest*.

Praise for *The Lost Elms*

'Her enthusiasm is contagious. I was inspired to download the Woodland Trust tree-ID app and resolve to pay more attention to our ligneous friends.'
Guardian

'Not just an elegy to our lost elms but also a meditation on life, culture and trees.'
Fred Pearce, author of *Fallout: A Journey Through the Nuclear Age*

'Haggith's captivating book is . . . engagingly written and has important things to say about globalisation, the threat of climate change and the value of biosecurity.'
Independent

'This book defies us not to fall in love with elm trees, with the idea of elms and all that their loss and what remains represents to us.'
Kirsteen Bell, *Caught by the River*

'Unofficial poet laureate of our woodlands.'
The Scotsman

'An engaging and inspirational portrayal of this culturally significant and iconic tree.'
Permaculture Magazine

Also by Mandy Haggith

Fiction

The Last Bear

Bear Witness

The Walrus Mutterer

The Amber Seeker

The Lyre Dancers

Poetry

Briny

Why the Sky is Far Away

A-B-Tree

Into the Forest

Castings

letting light in

Earth Wonderings

Non-fiction

Paper Trails: From Trees to Trash –
The True Cost of Paper

For Bill

Contents

One – Death: Introduction   1

Two – Life: Ecology   22

Three – Death: Dutch Elm Disease   49

Four – Life: Healing Uses   74

Five – Death: Coffins and Cartwheels   98

Six – Life: Elms Around The World   124

Seven – Death: European Elms and their Folklore   152

Eight – Life: American Elms   176

Nine – Death: Elms in the Arts   205

Ten – Life: Lessons from the Elms   241

'Elm tree talk' 271

The Elm Family 278

Suggested Further Reading 282

Acknowledgements 283

References 285

Chapter One

Death: Introduction

I grew up in an elm wood, one field away from our semi-detached brick house in a cul-de-sac at the top edge of a Northumbrian village. Throughout my childhood – during weekends, on summer evenings and after school – I frequented what we called the bluebell woods, a riverine forest flanking the banks of a tributary of the River Tyne. The west bank was Mirkwood, a conifer plantation, but the east bank, the side closest to the village, was a native woodland full of magnificent elm trees, grand structures that could hold up a cathedral roof. I scuffed about among them, making dens with other children from my neighbourhood, gathering sweet-smelling herbs like woodruff to dry under my bed, looking for fox dens, caterpillars and sticklebacks, sitting quietly at dusk until badgers emerged from their sett. Elms shaped the sacred playground of my childhood.

But then they started dying.

Dutch elm disease, a ruthless killer, had arrived. Looking back, it felt quick. The canopy thinned. The trees began to disintegrate as limbs, bare all summer long, became brittle and snapped in winter gales. What had been a closed, green world became inhabited by ghosts.

The woodland was on ground too steep for cattle or tractors, thus spared the agriculture that otherwise surrounded our village. It swathed a gully gouged out by a stream in its rush to the Tyne, and its luxuriance spread along the north bank of one of the river's many meanders. The woods connected the willowy valley floor with moorland up above it, forming a corridor for wildlife, dog walkers and feral children through an otherwise intensively farmed landscape.

One winter not long after the elms had died, the riverbank, no longer held together by a living structure of interlocking roots, collapsed during extreme rain. The resulting landslide blocked the road along the riverside for the next two years. During this time, I started at the middle school just a couple of miles away in the next village upriver. Before the slip, we could have walked or cycled along to school, but now we had a ten-mile bus journey involving a detour up out of the valley and then back down again further along.

As part of the road repairs, the fallen elm trees were cut away. Stumps and bare earth were all that remained. What had been the heart of a woodland thronging with life was now suddenly its edge, a precipitous drop to the seething river below. As I grew up, that wounded land became a place I loitered in. At the age of eleven I won a place at a school in Newcastle upon Tyne, something I was supposed to be pleased about, not least because my elder sister and brother both already attended. But my friends in the village went to the nearby comprehensive school, which I would have much preferred, and I was quite reasonably shunned for going to the 'snob school' in the city. I was often lonely, and the ghost-filled woods became my solitary haunt. What had once been a place of childhood certainty and comfort, a place to gather and play, was now devastated. The local kids all went elsewhere – down at the riverside among the ever-vibrant willow thickets, along the old railway line or around the allotments and the playing field – but I no longer felt welcome to be with them. In the woods I could avoid the risk of bumping into others: the trees wouldn't judge or shun me.

At the top of the landslip, at what was now the edge of the woods, with a fine vantage point over the river and of any approaching threats, one of the elm stumps became *my* stump, my refuge. I visited repeatedly to think my philosophical and mostly angst-ridden teenage thoughts, to smoke my illicit cigarettes while daydreaming about rock star liaisons or unlikely future careers. When there was difficulty at home, I felt safe alone in that scarred space. My stump, a round plate more than a hundred rings in diameter, was easily wide enough to sit on with my legs outstretched, but mostly I'd perch on the edge, kicking my heels on what remained of the trunk. Meanwhile, this great tree bole taught me its lessons, first and foremost that nature is so obviously vulnerable and that life as we see and live it is precarious.

The elms taught me that ecosystems are not closed worlds, and the underlying forces on which they depend – weather, soil and even rocks – are all dynamic. This was the 1970s – fear of a nuclear winter, with a detonation potentially triggering an apocalypse, pervaded our collective psyche. I learned in those times about the 'butterfly effect': how in a complex system, a tiny perturbation in one place and time can be magnified, with a ripple transmitted in ever-multiplying form until it becomes first a wave and then a tsunami. A beautiful flutter can snowball into devastation. In my home woodland it wasn't the wingbeat of a butterfly but that of a beetle that had caused chaos, and I had proof beneath my feet that one tiny insect could precipitate a landslide.

The beetle in question is *Scolytus scolytus*, the so-called large elm bark beetle, which is actually only about half a centimetre long, a shiny little creature with cherry-brown wing-cases, a polished black thorax and furry head. In itself it's an innocent enough little borer, and a big elm tree could survive its nibbles without too much difficulty, but its killer impact is as a carrier of infection, specifically the fungus *Ophiostoma novo-ulmi*, which causes Dutch elm disease.

There are several confusing things about the nationalities in this story: the English elms, which have died in their millions of the disease since the 1970s, weren't really English at all – they were field elms brought to Britain from further south in Europe by people in earlier millennia. Nor is the fungus in any way Dutch. It arrived in Europe from Canada or possibly the USA, though it isn't native to either; it had earlier European relatives probably originating somewhere in Asia, and in what sense would a fungus have a nationality anyway? The disease got its name simply by being identified by three female Dutch epidemiologists a century ago.

The beetle's habit of penetrating under the bark to feed transmits the fungus into the sapwood of the trees, with devastating effect. Death is rapid and almost inevitable. Since the current form of the disease arrived in the south of Britain in the late 1960s, it has gradually moved north, wiping out at least 25 million elm trees. In the USA the toll is more than 100 million elms. The blight is now globally spread by several species of bark-boring beetle, killing elms across Eurasia and America.

This book explores why this matters: the huge significance of losing these trees that have played such a large role in our cultures, as symbols of freedom, of death and the afterlife, of memory and of survival. The story of their battle with Dutch elm disease – from the calamity of its unchecked spread to the careful efforts being made to develop resistance – holds vital lessons for the similar challenges that animals and others trees will surely face due to climate change. Although there are many elm species, I often refer to the whole family collectively as 'the elm', and examining its ecology, practical uses, folklore, culture and literature can reveal why the tree is so intimately entwined with our history. By grasping what this iconic tree taught our ancestors, we might just reveal its one present message to humankind. The elm's fate hangs in the balance, and its survival will be decided by our effectiveness in limiting and adapting to climate change. This is ultimately a story of hope, born from my reverence for both the lost and surviving trees, to show that even in its darkest hour, the elm can be a symbol of the resilience we so urgently need.

In my early twenties I moved to Scotland and eventually found myself in the Highlands. Over the next two and a half decades, I watched Dutch elm disease steadily progress northwards towards me. A key fact about the *Scolytus scolytus* beetle is that only on a warm summer day will it venture out of its woody burrow to seek an alternative tree to chew on. New generations of beetles that hatch into overcrowded trunks will find their efforts to expand their territory thwarted by the cold. There is thus a clear line across Russia, Canada and European countries, an isotherm, north of which is a no-go area for adventuring beetles, too chilly for survival. Here in the UK, the dreich and nippy climate of the Highlands of Scotland has prevented not only patio dining, beach bikinis and cultivation of more-than-once-in-a-blue-moon pears; it has also put off the otherwise inexorable movement of Dutch elm disease. The wych elms or Scottish elms (*Ulmus glabra*) are Britain's only native elms, here since the retreat of glaciers after the last ice age, but thankfully northern Scotland, though rich in midges, has tended not to have elm bark beetles. Global warming, however, is changing this – each year that crucial isotherm shifts north and west and the beetles' domain expands.

When I first moved to the Highlands I lived in a hamlet called Struy, up Strathglass. Downstream is the town of Beauly, with its excellent greengrocer, hardware store, artisanal shops, cafés and Indian takeaway. It's an attractive place with a big central square flanked on the west side by an ancient priory founded in the thirteenth century by a Valliscaulian religious order, leading to the town's Gaelic name, A' Mhanachainn, or 'the monks'. The monk order's French origins probably account for its English name, from the French *beau lieu* (beautiful place), though this is apocryphally attributed to a remark by the French-speaking Mary Queen of Scots.

For centuries the priory gates were guarded by a monumental tree, Europe's oldest elm. It was planted in the 1200s when the church was built. We know this from the monks' documents and subsequent plans, which show a clear double line of elm trees forming a shady avenue in the approach to the priory. By 1998, when I first met it, the gateway tree was the last one standing: a vast ent-like tree, its trunk so broad it would need several tree-huggers to stretch arms around it, a commanding presence in Beauly's central square. The community was rightly proud of being host to Europe's oldest elm, particularly when the continent's populations of the tree were decimated by disease.

The Beauly Elm stood as a symbol of survival until the early 2020s, when it became clear that it was battling infection and losing the fight. In spring 2022 it put out a few leaves, but as these withered a local artist, Isabel McLeish, formed a plan to commemorate the life and wonder of this marvellous specimen. Together with the arts organisation Circus Arts, she coordinated a community outpouring of feeling for the tree, with people of all ages writing and creating art to honour its passing. I was commissioned to write about my own close encounters and intimate moments with the dying elm.

When I began, I expected to write a piece of mourning, an elegy. The project coincided with my father becoming increasingly frail after my mother's death and I assumed that my writing would meditate upon senescence and decline. Instead, I found myself marvelling at the tree and I ended up writing a celebration in the voice of the elm itself. For me, the elm was not a single organism but a vibrant living community – from the epiphytes in its crown to the fungal hordes in its roots, from visiting birds to resident wasps, spiders and, of course, beetles. The tree seemed to have much to say about the interdependence of species and about the importance of engaging all our senses and cultural values in wonder. The result is a poem for performance called 'Elm Tree Talk', with an accompanying set of 'rootnotes' for the version in the commemorative publication *Guardian of the Gateway: 800 Years of the Beauly Wych Elm* (see Appendix 1).

Circus Arts' activities culminated in September 2022 in a day of performances and participatory art with a milling crowd of local people sharing stories of what the elm meant to them. Just as the tree was a community of many species, it had connected the human community that has lived alongside it, providing a tangible link to our history through the power of its longevity. Its sheer age gave it profound value. Because trees can transcend our generational limitations, we entangle them with stories and thus they become cultural repositories and storehouses of knowledge.

Elms have a similar cultural significance around much of the world, as there are few places that don't have one species or other. They are an ancient family of trees, with many different species across the temperate Northern Hemisphere and down into the subtropics and tropics of Asia and South America. Their shapely forms mean that, as well as growing wild, they have been deliberately planted in their millions. The monks in Beauly knew what they were doing when they planted an avenue of elms to grace their beautiful holy place: two parallel lines of elms create a church-like nave, an arch-shaped cloister that draws the eye towards the temple at its end. Town and city designers the world over, from Scandinavia to Australia, have used this effect to give a sense of distance or simply to create welcome summer shade and winter shelter.

Tunnels of American elm were a fundamental part of urban design in many cities in the USA until they were devastated by Dutch elm disease in the mid- to late-twentieth century, and the iconic 'before and after' photographs of these cities show just how heartbreaking the disease must have been to millions of people. American tree geneticist David Karnosky's paper about the disease featured an image of Cornell University, with quadrangles of lawn lined by elm archways, a gorgeous shady environment in which students and scholars strolled at ease.[1] After the trees died, it became a stark place, windswept and spartan, with only the severe lines of buildings to frame its learning. A photo by Jack Barger, in communications professor Philip Hutchison's paper about journalistic coverage of the 'elm blight', shows a road in Detroit that seems broad and spacious, with houses offered privacy and shelter by trees, cars parked in dappled tree-light and a circular glow at the end of the tunnel somehow suggesting the optimism of the American dream, contrasted with the later, now treeless, street, which is purely functional, dominated by asphalt and completely lacking in privacy.[2] As well as the aesthetic impacts, there were significant financial losses: the absence of the elms' cooling shade in summer and prevention of wind-chill in winter, plus a reduction in air quality, have been accounted as being worth billions or even hundreds of billions of dollars.[3] In many wine-growing regions, such as France and Italy, elms were used to support and shelter vines, and again their losses have been both culturally and economically damaging.

People form strong emotional bonds with trees. In John Miller's book about the cultural significance of forests, *The Heart of the Forest*, he vividly describes the huge social uprising in Sheffield when the local government set about felling thousands of street trees. Emotions often run high when neighbourhood trees are threatened, and the felling of a single garden tree can be soul-destroying for the family who picnicked beneath it, climbed its branches or just enjoyed its calm presence through the changing seasons. During the COVID-19 pandemic, I partnered with Scottish Forestry to research how people feel about trees in the landscape. Participants poured out powerful expressions of how crucial they were finding the presence of trees close to their homes, as living beings, custodians of nature, sources of the very oxygen we need to breathe and, in a phrase used by one person, which became the title of a collective poem that went (modestly) viral on the internet during lockdown, 'The One Thing We Can Hug'.

Since 2011 I have been running a project called A-B-Tree (A-B-Craobh in Gaelic), inspired by an ancient connection between trees and the written word: in the Gaelic-speaking parts of Scotland and Ireland, a native woodland species is linked to each letter of the alphabet. Known as the Gaelic Tree Alphabet, these links have been around for at least 1,500 years, maybe longer. A-B-Tree shares folklore, poetry and practical and ecological knowledge about the trees, and despite running for more than a decade, the project is far from finished – there is so much tree lore, so much wisdom and traditional knowledge, so many stories and ways that trees have been useful or significant, so much still to learn. In some versions of the Gaelic Tree Alphabet (there's a debate, as we'll see in chapter seven), the letter A stands for the wych elm: each letter is the first of a tree name in old Gaelic, and the name 'ailm' is ascribed to elm. The Gaelic Tree Alphabet is just one example of the wealth of elm's cultural connotations and, as later chapters of this book will explore, the lore about elm is as broad as its trunk and as deep as its roots.

Most famously, elm is a tree associated with spirit realms and passages to the afterlife. Whether as a portal to a fairy land, home of a gnome or a stage set for a spiritual ceremony, the elm tree has always been seen as a doorway to other worlds. It would have been a hugely significant choice for monks in ancient times to line their walkway to their holiest place. The Guardian of the Gateway has watched many coffins carried under it on the journey to a grave and whatever follows after. Many of those coffins may have actually been made of elm as it is a favoured material for housing the dead.

Elm also has healing properties – its roots were used to knit bones, its leaves for dressing wounds, its bark to ease gut problems. It has many other practical uses: as well as coffins, elm wood has long been valued for making tools, wheels, boats and even water pipes. This is a tree deeply rooted not only in our lands but in our cultures. No wonder its loss from our woodlands, hedgerows and town centres has left us feeling bereft, though I'd argue we need to do much more to respond to this feeling.

Although treated as an honorary Beaulian while the Guardian of the Gateway celebration took place in 2022, for the past twenty-four years I have lived a hundred miles north-west of there. Uniquely in Britain, old elms are still flourishing near my home along with their assemblage of other species, so far unaffected by Dutch elm disease. I live in a zone of hope, beyond the fringe of the beetle's range. This isn't, of course, any kind of guarantee – climate change is shifting the temperature contours northwards; the isotherms on the climate maps are redrawn, little by little, with every year that passes; and the beetles follow. The line between intolerable and hospitable is drawing nearer. So far, we're still cool enough, just, to put the bark beetles off.

I live on a woodland croft on the rocky shore of a sea loch. It's a miraculously beautiful environment with wild weather and plentiful biodiversity. Among the twenty-three species of tree in our village there are wych elms. They are not as common as birch, hazel, aspen and willow, but some of them are mighty specimens. We find them teetering in craggy and inaccessible places or gracing our woods with their unique palette of green and gold. Efforts are underway to propagate trees that show resistance to the disease and plant them out to increase our woods' genetic diversity. This will hopefully ensure all the other species that depend on elm have some to live on for perpetuity.

The elms have come to seem to me a symbol of the quandaries we face in trying to address climate change. We have pumped so much greenhouse gas into our atmosphere that global warming is guaranteed. It's no longer a question of whether we will warm further, but how much and how fast. How will we run our society, and what changes are we each personally willing to undergo? How we manage this transition will also have huge impacts on how well we adapt to future climatic conditions. How we live and work with our surviving elms in north-west Scotland is a microcosm of these wider questions. It's unlikely we can stop the beetles' progress, but can the remaining elms somehow be protected from death by fungus? I want to believe they can.

We're liable to take nature and the people around us for granted: it wasn't until my mother was dying, and then very soon dead of cancer in 2016, that I appreciated the foundational role she played in my life and in my identity. Prior to her death I had visited her and my father perhaps once a year, seeing them at family gatherings, arts festivals or when they visited me, but the crisis of Mum's death and the subsequent need for me and my siblings to care for our father meant that I was suddenly making much more frequent trips to Northumberland and staying there for longer periods than I had for decades. Although the village no longer felt like home, amid all the emotional turmoil the woods were still the refuge they had always been, and I took great comfort in the way the scars in the landscape that had been caused by the loss of the elms had healed over, with oaks, hollies and hazels filling the gaps. Although my mother was no longer with us in bodily form, her presence still felt strong, her significance and guidance still active. I woke from dreams of hugging her, heard her voice giving me good advice, made imaginary phone calls to discuss dilemmas and found that she was still able to resolve them, and I was better able to heed her wise recommendations than I ever had while she was alive.

Mourning is a hugely powerful process. Thanks to support from my sister, in particular, I was able to dive deeply into it and allow it to shape me over the years that followed. I did this grieving in private, however, because the processing of loss does not feature frequently in our culture – that famous British stiff upper lip. When our village lost its elm wood, perhaps we should have had a public outpouring of sorrow, rather than simply getting the bulldozers in to build a new road. I believe our insufficient attention to environmental grief, the lack of mourning for the other life forms we share this planet with, prevents us working hard enough to stave off further losses. Tree disease prevention methods urged by scientists have often fallen on deaf ears; it is politically acceptable to shrug to the loss of trees, and indeed to climate change in general. We know so much, yet our emotional blind spot risks making us incapable of acting sufficiently to avert a crisis. I hope that by focusing on the connections between ourselves and the rest of nature, we'll be better able to manage the grief of the losses we have already incurred and change our ways of being in the future.

My local elm wood's collapse was the start of my awakening from childhood and loss of innocence. Can the elm species' global predicament wake humanity up to our place in the complex spiral of life, to jolt us out of complacency before it's too late? Elms are more than symbols of death, and their story can be greater than just one of loss. Their against-all-odds survival can act as a spell to teach us to value the rest of nature, and theirs is, above all, a tale of burgeoning hope rooted in resilience.

This book's ten chapters alternate between themes of death and life. The next two chapters address elm ecology, first by looking at the evolution and global distribution of the tree, and then delving into those ecological relationships that cause Dutch elm disease. The following three chapters turn to elm's importance to human societies over the ages – the many practical ways in which the elm has supported human life through its leaves, flowers, fruits and sap; the traditional use of elm wood; and a closer look at elm's fibrous bark, including the story of the world's earliest paper, made from the bark of the qing tan tree, a Chinese elm relative. We will then explore the lore and stories associated with elms in Europe, which are rich in linkages with death and the afterlife; the elm's role in national identities in North America and how it has become the tree of liberty; and how Canada and the USA have tried to treat infections by Dutch elm disease and limit its spread. The penultimate chapter explores how elms have been represented in and inspired art and literature around the world. The final chapter looks forwards to consider the possible fates of elms, exploring how botanists and foresters are working to develop resilience in elm tree populations and protect other species that depend upon these beautiful trees. Here lies hope.

At the end of each chapter there is a vignette of a particular tree. In the 'death' chapters, these commemorate historically important elms, while the 'life' chapters feature meditations on visits to particular trees in my home parish of Assynt in the north-west Highlands of Scotland.

The Washington Elm

Trees are essential to our survival, as food sources, as protectors of water courses and as climate-changing gas absorbers among other things. While forestry work to provide wood products is necessary, it does not follow that all trees are replaceable or that no tree is individually significant. On the contrary, many individual trees have shaped our cultures and civilisations. Most religions hold particular trees close to their hearts, such as the aspen used for the cross on which Jesus Christ was crucified or the banyan tree under which the Buddha achieved enlightenment. Many nations have an iconic tree where a crucial historical event took place, and this tree and its story can end up playing a key role in that country's national identity. The globally iconic Washington Elm is one such tree.

Legend has it that on 3 July 1775, General George Washington, aged forty-three, took control of the Continental Army under the shade of the spreading branches of the elm tree on the Common in Cambridge, Massachusetts. Almost exactly one year later, Congress voted for independence of the Thirteen Colonies from Britain, forming the United States of America. George Washington led the army to liberation through the American War of Independence, and in 1789 he became the USA's first president. He is often referred to as the 'Father of His Country'. The elm tree under which he is said to have assumed military command lived on long after his death in 1799. Paintings and engravings show it to be an elegant specimen, with a sturdy, handsome trunk several metres tall, branching out into a ring of upwards-reaching limbs forming the classic vase shape for which American elms are so famous. A plaque at its base read: 'Under this tree Washington first took command of the American army, July 3rd 1775.'

The tree didn't really come to national prominence until many decades later. In 1861 the American Civil War was looming and a huge crowd gathered around the elm, apparently unorganised, to discuss how to protect the Union. The tree was decorated with Massachusetts state symbols, the Star-Spangled Banner and all manner of shields and signs, and it was subsequently used in propaganda as a symbol of all that New Englanders held dear about their united nation. It was central to the community and became the preferred venue for launches of new organisations and initiatives, such as the Harvard University baseball club, and the terminus of an annual parade through Cambridge on the Fourth of July. It was an essential stop on the itinerary of foreign dignitaries, and by the start of the twentieth century it had become one of Massachusetts' primary tourist destinations.

The tree died in the summer of 1923, most likely poisoned by traffic fumes. Analysis of its tree rings showed that it was between 204 and 210 years old, which would have made it at least 50 years old, a mature, though not huge, tree, when Washington formed his army. When it finally toppled, there was a frenzy as people tried to secure a twig or a fragment of bark or a piece of its wood. The scenes of axe- and saw-wielding souvenir hunters were so chaotic that the city government had to step in to stop a riot. It then undertook an international distribution programme of the wood with the explicit aim, according to historian Samuel Batchelder, of making the loss of the tree 'an object lesson in patriotism for the whole country'. The careful placing of more than a thousand fragments of the Washington Elm across the USA and beyond ensured the tree achieved a standing close to that of a religious relic: each of the forty-eight states at the time received its chunk of timber with a request to do something special with it; a cross-section of the trunk showing all of its rings was placed in Washington's former residence at Mount Vernon; artists carved works of exquisite complexity for public display such as book covers for libraries and picture frames for galleries; a gavel was made for each legislative body in the country; and pieces were presented as formal gifts to allied countries through foreign ambassadors. Pageants and ceremonies at the site of the tree commemorated the events of 1775.

The Washington Elm now has its own significant descendants. In 1902 a scion of the famous tree was taken from Massachusetts all the way to Seattle, in Washington state, and planted in the University of Washington grounds by a history professor named Edmund Meany and other dignitaries in a grand patriotic ceremony. In the years following the death of the Washington Elm in Cambridge, the university campus gardener, Ludwig Metzger, set about propagating from the precious daughter that was in his grounds, and these grandchildren were likewise distributed around the country, including one that was returned to the site where its famous grandfather grew on the Cambridge Common. These progeny of the original Washington Elm cross many people's paths, including mine. For example, one is down by the harbour in Bellingham, the town closest to where my brother now lives, and my nephew studied at Washington State University, where that daughter elm still stands.

In the USA, some people trace their family roots to the American War of Independence, allowing them to join the Sons or Daughters of the American Revolution. Local chapters of these national patriotic organisations have perpetuated and protected the Washington Elm's lineage. The Daughters of the American Revolution were particularly active in this process, planting and tending to the trees in their new homes, and to this day, the organisation has many chapters with elm trees on their agenda. A Colorado chapter has been working on micropropagation techniques to keep the Washington Elm descendants coming. The Michael Trebert Chapter has been running an American Elm Heritage Project in the lead-up to the 250th anniversary of American independence in 2026, and in 2022 they led a commemorative planting of American elm saplings in Forks, Washington. They assert 'the importance of reestablishing American elms in our nation' as symbols of liberty.[4] The lineages of these people and those of the Washington Elm tree are mutually supportive, in a fascinating blend of the significance of family trees, as if when we protect a tree it, in turn, shelters our history.

This symbolism of the Washington Elm and its offspring is hugely compelling, especially because its story is at least partly myth. There was most likely a throng of military people on the Cambridge Common on 3 July 1775, and George Washington was certainly in the area. Around that time he did take effective control of the army. But what is less clear is the relevance of a fifty-something-year-old elm tree, although records show that the weather was poor and it would have offered shelter in the days before Gore-Tex and rubber boots. But there's really no hard evidence that any ceremonial event actually happened there. A diary entry by Dorothy Dudley refers to the formal taking of command as happening under an elm tree and being 'a magnificent sight', but this diary was later revealed to be a literary forgery, a fiction written by author Mary Williams Greely.

Does the lack of historical evidence matter? The meaning of the tree as a representation of freedom from colonial control has grown powerfully in the intervening centuries, as we'll see in chapter eight. In 2025, descendants of the Washington Elm will be ceremonially planted as 'liberty trees' to commemorate 250 years of independence, demonstrating a power that transcends its shaky foundations in documented history. Symbols and stories are powerful because of the meaning we imbue them with, not just their link to real events. The Washington Elm reveals how myths are created and come to influence our societies. Myths are being made all the time, at all levels – few have national or international significance, but we are all subject to them. In the fallout following my own parents' deaths, I became acutely aware of the stories that our family had told, and re-told, for decades, despite their tenuous roots in reality. Every person, family and nation must find their own myths to bind them together and hold them dear. Sometimes we are so protective of these stories that any attempt to challenge them are met with horror or even fury. This is exactly what happened when Samuel Francis Batchelder argued that George Washington had not, actually, taken control of the army under the elm tree – debate around this raged.[5] Those for whom the myth had meaning felt deeply threatened by a truth that could undermine it. Regardless, the University of Washington's managers have no intention of removing the plaque from their tree – for them, the story is what matters.

The Washington Elm is just one example of the many ways in which elm trees don't just stand – they stand *for* something. I'm not alone in having grown up with them as a significant and meaningful part of the landscape of home, the loss of which has reverberated through ensuing decades. At the position of the original Washington Elm, its grandchild tree still stands, on Garden Street across from the Radcliffe Institute for Advanced Study at Harvard University. It's a splendid tree, with a broad crown forming a perfect umbrella curve over a characterful network of forked branches. A circular low metal fence ensures that its inner roots are saved from trampling. Long may it stand proudly with its message of freedom.

Chapter Two

Life: Ecology

There are many different kinds of elms, globally, and they have been categorised and re-categorised several times. For mere mortals like me, taxonomy is a dark art. My mother was a biologist who did her PhD on frog genetics but whose real love was plants, especially ferns. She would make long lists of plant species she had spotted and always had an appropriate field guide, wherever in the world she might be. From an early age she made sure I had a copy of *The Pocket Guide to Wild Flowers* by David McClintock and R. S. R. Fitter, which she considered the definitive field guide to flowering plants, and when I drowned my copy on a camping trip to Ireland in my early twenties she scoured second-hand bookshops until she secured me another copy. She bemoaned the changes in taxonomic systems that placed plants she was familiar with into new families or that created distinct genera where one genus used to do just fine, because that meant that her favourite books might not be strictly correct anymore, but she also fiercely defended the rationales behind such changes as scientific advances, small victories for the march of knowledge towards a fuller understanding of our fellow species in the natural world. However, I'm no geneticist, and my botanical knowledge is strictly amateur – born of love, as the Latin origin of the word suggests, but not professional.

The tree is not only a symbol of life but a metaphor for the branching structure of evolution, akin to the family trees we use to track our personal genealogies. A taxonomic tree places species as the leaves, the groups of similar species or genera as twigs, more broadly related families as branches and so on. Most, though not all, modern plant taxonomy is based on analysis of the molecular structure of the genetic material – the DNA – that is inherited from one generation of a species to the next, gradually mutating over time. This enables, in theory, a precise definition of which plants are close relatives, which have common ancestors and so on. In the eighteenth century, when Swedish biologist Carl Linnaeus first proposed a systematic approach to the naming of species based on their family resemblances, all that botanists had to go on were the external forms of plants, and as a result plants were gathered together into families and genera based on similar flowers, fruits, leaves, branch structures, bark texture, etc. As the centuries went by, the criteria for grouping plants into the same genus were refined and argued over, and elms have been particularly ambiguous. There are plenty of trees that look like elms, have elm-like reproductive parts, behave like elms and in particular have very similar fibrous bark, but they turn out to have rather different DNA so have been expelled from the elm family. The debate about how exactly to classify elms has been so fraught that a name has been coined for the botanists who indulge in it: pteleologists, from the Greek for elm, *ptelea* (πτελέα), a pun on teleologists, who study the ultimate purposes of natural objects.

A classic work of classification by the French botanist Charles-François Brisseau de Mirbel, his *Elements of Plant Physiology and Botany*, published in 1815, grouped together many elm-like trees as the family known to this day as Ulmaceae, including many trees known as *Celtis* (hackberries and nettle trees), *Pteroceltis* (of which there is just one species, *Pteroceltis tatarinowii*, the blue sandalwood or qing tan tree) and others, which spent the next 150 years being referred to as elms, catalogued together in herbaria and planted in botanical gardens in collections of elms, but in recent decades they have been turfed out of the elm family due to their DNA and grouped instead in the Cannabaceae, with nettles and hemps. This was significant in China, where qing tan tree bark fibre has a particularly interesting story, as we'll see in chapter six. These days it has to make do with being just an honorary elm.

At the most general level, the elms are a family of trees called Ulmaceae, as classified in the *Plants* journal[1] by Swiss botanist Yann Fragnière, Chinese botanist Yi-Gang Song and their colleagues. Although there is a tendency to think of elms as 'northern' or temperate trees, perhaps due to their relative abundance in more northerly landscapes, the diversity of the family is greatest in the subtropics. It consists of seven genera, three of which are tropical (*Ampelocera*, *Holoptelea* and *Phyllostylon*) with the other four temperate or subtropical (*Hemiptelea*, *Planera*, *Ulmus* and *Zelkova*), consisting of a total of fifty-six species. The tropical genera have thirteen species in total, so the majority of elms are temperate or subtropical, and the *Ulmus* genus accounts for more than half of them – currently thirty-five different elms are counted. China has the greatest diversity of elms, with twelve different species across three genera. South-eastern USA is another elm-diversity hotspot, with six species in Arkansas alone.

In 2021 a classification of the elm genus was published by Alan Whittemore of the United States National Arboretum in Washington, DC, and several academic, arboretum and field museum colleagues in Chicago. They break *Ulmus* into three subgenera: the *Indoptelea*, which contains a single species, *Ulmus villosa* (the Marn or cherry-bark elm); the *Oreoptelea*, which divides into two sections covering nine more species, including *Ulmus americana* (the American or White elm); then all of the other elms, twenty-four species of them, with numerous subspecies and varieties, which are grouped into the subgenus with the same name as the genus overall, *Ulmus*, which is in turn divided into five sections. One of these sections also carries the genus name, as if these elms are the elmiest of them all, and it contains 'my' elm, *Ulmus glabra* (the wych elm). The so-called 'English elm' is really a field elm, *Ulmus minor*, often clarified as the cultivated variety 'Atinia' because it is believed to have originally been introduced to Britain from that part of Italy. It is placed in the section *Foliaceae* of the *Ulmus* subgenus. There are elms named for every place you can think of in the Northern Hemisphere – Japanese, Himalayan, Chinese, Manchurian, Tibetan, Siberian, Mediterranean and Mexican, including some highly specific places, such as the Hangzhou or Harbin elms, and some named for form, like the cherry-leafed or the long raceme or large-fruited elms. There are those whose stories beg to be investigated, like Father David's elm (named after a Basque Catholic priest, Armaud David, who found it in China) and the September elm (so-called for its habit of autumn flowering), and there are some, like *Ulmus chumlia*, with no common name at all in English, though the people of the Himalayan midhills of Nepal and Kashmir who cut it for animal fodder no doubt have their own names for it.

Ulmus chumlia can give us a little flavour of the name games played by plant taxonomists. In 1848 it was known as *Ulmus virgata*. By 1919 it was deemed a variety of *Ulmus pumila* or by another botanist as a variety *subhirsuta* of a species called *Ulmus wilsoniana*. For a while it regained *virgata* as its variety name, but of a different species *Ulmus × androssowii*, before being re-classed in 1991 as the variety, *subhirsuta*, of that species. Are you still following? Of all these names, only *Ulmus pumila* is still in use, but these days it's the scientific name of the more northerly Siberian elm. No wonder it's easy to get confused by elm taxonomy! Suffice to say that elms are a much argued-over group of trees among scientists and keep being reclassified in different ways as we learn more about them.

The subject is further complicated because there are almost as many species of extinct elms as there are living species – and although this is all part of the natural order of things, it is a salutary reminder of their, and our, vulnerability. We know about the extinct elms from the fossil record, which goes back to the early Paleocene Epoch, around 66 million years ago, by which time elms started cropping up all around the Northern Hemisphere. So elms have been gracing Earth since the dinosaurs roamed the planet. The oldest elm fossils of still existing genera date back to the early Eocene, around 50 million years ago, in China. Some not-quite-so-old elm fossils remain of existing species which lived in North America around 10 million years later. It's easy to bandy figures like this around, but to actually imagine something enduring on this planet for 40 or 50 million years is mind-boggling. Compared with our human lifespans of a mere eighty years or so, or the few tens of thousands of years of human presence in Europe, elms have been here for so much longer. Even the entire existence of our species takes us back only a tiny fraction of the period that elms have been rooting and fruiting and shooting the breeze.

I was first struck by this temporal vastness at the Royal Botanic Garden Edinburgh when I was their poet in residence in 2013. I had been there a couple of days when I strolled out of the palm house chewing words over in my head and stumbled to a halt, finding my way blocked by a strange line of brown stones shaped like chunks of a tree trunk. Having literally tripped over a fossilised *Pitus withami* tree of some hundreds of millions of years of age, I started noticing information boards with similar phone-number-length dates on them. With my head starting to spin, I came across a quote from the nineteenth-century natural philosopher, John Playfair, who said that his 'mind seemed to grow giddy by looking so far into the abyss of time'. I, too, felt like I was teetering on the edge of an unfathomable deep. It is no surprise that so many cultures have trees as symbols of life itself, or believe that humans originated from them, when an individual tree can outlive a dozen human generations, and its forebears have been striding the continents since long before anything remotely like humans were even imaginable. I find the longevity of plant species to be a source of profound hope for the planet: whatever the mess we make of this place during our tenure as Earth's most meddlesome species, there will surely still be trees and other plants to heal over the scars we leave. I once visited an Inca city in Peru, which in only a few hundred years had reverted from a metropolis of thousands of urban people to magnificent forest with vast trees rooted through grand stone buildings, creepers and undergrowth burying most signs of human life under a riot of green. Given that elms cover so many different habitats on Earth, from the tropics to the tundra, surely there will be somewhere for them to carry on; most elm species prefer moist places, so they'll find a space as long as there's something between desert and ice.

One of the most astounding places elms grow is on Canna, a remote Hebridean island with a valuable safe harbour for sailors (my summer obsession). There is nothing to the south-west of Canna except a few thousand miles of Atlantic Ocean, and that's the prevailing wind direction, scouring the island with salt-laden rain and gales. Yet here I found a craggy slope swathed in elms, sculpted by the wind into wedges, repeatedly clipped to the distinctive wind-raked shape of the slope. In mid-July, the trees were in full leaf and thriving, and underneath the triangular canopy the ancient trunks were chunky and strong. They must have been hunkering there in the teeth of the wind for centuries. Here was a perfect example of elm's toughness and ability to survive the ravages of severe weather.

Elms need to be tough because unfortunately their conservation status is grave, with a third of all elm species assessed by the International Union for Conservation of Nature (IUCN) as being on their Red List as 'under threat'. Six species are endangered, including *Ulmus americana* (American elm). Two of these are critically endangered, described as 'on the brink of extinction' by Fragnière: *Ulmus gaussenii* (the Anhui or hairy elm) from eastern China and *Zelkova sicula*, a native of Sicily. Only twenty-six adult trees of the Anhui elm are known in the wild, in a single patch of less than ten hectares. *Zelkova sicula* grows in the Iblei mountains in Sicily in stream valleys in just two locations. A similar endangered species, *Zelkova abelicea*, grows only on the island of Crete, where it suffers over-grazing from feral goats.

The zelkovas are the most endangered of the elms, with all six species under threat. They are a group of very attractive trees, which turn out to be close enough relatives to the other elms to be officially included in the family, though as a distinct genus they don't usually get called 'elms'. Zelkovas have very simple fruits, called achenes, that contain a single seed with a skin called a pericarp; they are deemed to be the most evolutionarily 'primitive' of the elms and as such offer an early insight in the elm's extraordinary evolutionary story.

The fruits of elms are one of its most distinctive features. From the zelkovas' simple beginning, elms have evolved various more sophisticated fruits. The thorn elm, *Hemiptelia davidii*, one of the East Asian species, which gets its scientific name from the same Armaud David responsible for Father David's elm, has grown a wing-like appendage on one side of its fruit, enabling it to be picked up and spun by a breeze. This evolutionary breakthrough is a feature of most current elm fruits. All of the 'true' *Ulmus* species have winged fruits, called samaras, in which the pericarp forms wings around the central seed, enabling them to be widely dispersed by the wind. Not every species relies only on the wind, however. Some of the other species, notably *Planera aquatica*, have evolved fleshy water-wings around the seeds so that they can float off to a new home down streams and rivers. The genus of tropical species called the *Ampelocera* grow colourful flesh between the seed and the pericarp, making them tasty to birds, which then excrete the seeds, scattering them widely. On many of the temperate species, the fruits develop and become noticeable in early spring, before the leaves open, as the clusters of samaras with their papery wings gleam green-gold in the low-angled light of the early part of the year. Rather wonderfully their green tinge is because they contain chlorophyll, thus feeding on sunshine while they get the chance before the leaves open. These delicate wings are petal-like in some species so these sprays of fruit can understandably be mistaken for blossom.

But the elm's flowers are very different. They are perfect, though completely lacking in petals. Perfect is one of those botanical terms that means something unexpected: a perfect flower is hermaphrodite, having both male and female sexual characteristics. Isn't it pleasing to learn that botanists have such a liberal attitude towards gender fluidity!

Elms don't bother with the colourful petals that many flowers use to attract pollinators because their approach to sex is simply to stand out 'in the weather in their altogether', as my grandmother would have said, flourishing their naked sex organs and waiting for a breeze to do all that's required to distribute the pollen that the male parts are offering up from their anthers, hoping no doubt that the female parts get what they desire from the process, with one of those grains of pollen finding its way to the stigma and thence to one of the ovules within the ovary. Letting it all hang out in the wind and rain may not be everyone's cup of tea when it comes to sexual satisfaction, but each to their own.

The lack of pretty petals, or indeed any petals (corolla) or sepals (calyx) at all, doesn't prevent elm blossom from having its own special beauty, if you are willing to look closely. The flowers form in tightly packed bunches, each consisting of a flaring cup with lobes, often of a lovely pink, purple or red colour, in which the male parts – four or five white stamens with pollen-bearing anthers, often a striking crimson – stand erect like lollypops, ready to shake their dust-like pollen into the breeze. The cups contain the female parts, a little green ovary with short style and two-pronged stigma for catching a speck of that precious pollen. There are some extraordinary pictures of elm flowers, taken by Francis Principe-Gillespie using microphotography techniques, which reveal the detail of these sexual features beyond what is visible to the naked eye.[2] He has captured the anthers breaking open to expose their pollen, like thick sheep-fleece, while the receptive female stigma is like some kind of strange, but exquisitely beautiful, spiny sea cucumber, covered in a mass of bright pink rubbery fingers, called papillae.

An elm tree will produce a vast number of flowers, with the bunches forming straight out of its twigs, so that the tree is festooned with these little pompoms. In our landscapes, wych elm flowering gives the trees a pink blush in the early spring when most other trees are still in their winter dormancy, other than the hazels, whose yellow catkins, also wind-pollinated, are always ridiculously early.

The resulting plethora of seeds should lead to vast numbers of young elms springing up all around mature elms, but in reality their numbers are held deeply in check because the seedlings are tender and delicious to all grazing animals. As the forester H. L. Edlin puts it, 'A young elm seedling or sapling, in a wood that is open to the browsing of sheep, goats or ponies, has about as much chance of survival as a lettuce in a rabbit hutch!'[3] In my neck of the woods, we need to add deer to that list of browsers, as our various species of deer are the primary limiting factor to the regeneration of the trees. The classic English landscape role of elms in hedgerows is also explained by their susceptibility to herbivores. Here is Edlin again: 'Any shoots that arise out in the fields will soon be bitten back . . . and in fact the only safe place for a tender elm seedling or sucker shoot to grow is in the heart of a hawthorn hedge, where a barricade of thorns will ward off hungry livestock.'

The reproductive process of a mature elm tree, from flowers to seed dispersal, is fuelled by the great energy source of the plant – its leaves. Of course the various different species of elm have leaves of different characters and sizes but most of them are roughly oval in shape, with a central midrib, usually asymmetrical with the leaf growing further up the stem on one side than the other. The leaf texture is rough on many species – including the wych elm; despite its scientific name of *Ulmus glabra*, which means smooth, its leaves can be close to the texture of fine sandpaper – though there are some smooth-leaved elms, such as *Ulmus rubra* (American slippery elm) and *Ulmus parvifolia* (Chinese elm), which has positively shiny leaves. The leaves of most species have jagged edges, often irregular, sometimes having multiple points. One of the defining features is that the leaves form in groups, usually of five leaves on alternate sides of a shoot, with the biggest leaf at the centre and the other leaves decreasing in size moving away from the top, somewhat similarly to the lengths of fingers on a human hand. The shoots appear to be reaching out with their biggest leaf leading the twig into its growing space, like big hands outstretched, groping for light.

The twigs on which the leaves of elms grow are mostly rough, almost bristly, and brown. In the notches between each leaf stem and the twig there are little round pinkish buds, which will be the flowers next season. Next year's group of leaves will grow from a more pointed bud at the twig tip, characteristically almost at a right angle from the twig. Once the twig is a year or two old it becomes smooth, paler and greyish, and the bark remains smooth while twigs grow into branches. The bark of the trunk remains pale, with a brown tinge, until on the older wood it starts to become fissured with age, splitting to accommodate its broadening girth. Some species of elm, including the field elm and Dutch elm, have strongly textured bark with vertical grooves.

Before they achieve great age, most elm trees are stately, with their branches lifting at a pleasing shallow angle from the trunk, and they are often described as 'vase-like' in form, particularly the American elm, which is known for its elegant natural shape that led to its mass planting for ornamental purposes. Older trees may have huge barrel-like trunks. After shedding branches in storms or being cropped for animal fodder, as was common in the past in Europe and still happens in other parts of the world, their girth may become disproportionate to their crown, so that they appear like Tolkien's ents: chubby, gnarled sages with limb-like branches. If they have the chance they will age to adorable ancient or, as foresters say, 'over-mature' trees. Depending on species this can range from a mere fifty years, in the case of the fast-growing Chinese elm, or many hundreds for the slower-growing species like rock and wych elm, with the oldest known zelkovas reaching more than a thousand years old.

Some species of elms are prone to hollowing, and there were some very famous elms in England with substantial interiors. One of these, called the Crawley Elm, was apparently 61 feet (18 metres) in circumference with an interior 'room' of 10 metres, with a paved floor, a table and chairs, and a lockable door on the south side of the trunk. According to local stories, all kinds of things went on in there: at one point a woman made it her home and gave birth to a child in it, and at others it was the venue for local feasts and parties. It was cut down in 1935 as it was dying, presumably of Dutch elm disease. An elm in Hampstead, London, had a staircase inside it leading to a platform big enough to seat six people. A group of hollow elms in Essex, called the Old Maids, were smugglers' hideaways, according to local legend. In wild woods and forests in Russia and North America, and in Europe before we eliminated them from much of their former territory, old hollow elms no doubt host over-wintering bears sleeping out the cold and snow.

One of the marvels of trees is how much of them is invisible to us. Images of roots, such as the amazing drawings of several

European species of elms by Dutch scientists at Wageningen University, reveal that there is as much of a tree below ground as above, with a vast and complex fibrous network structure radiating out beneath the tree.[4] It is always worth stopping to ponder what's below the soil's surface, because in many ways it is the most vital part of the plant. Many trees, elms included, can survive being cut down to the ground, and will regrow from the bole that is left, usually with multiple trunks. The deliberate management of trees in this way is called coppicing. Even if not cut down, elms will put stems out from around the base of the tree, called suckers, and a tree under stress will often sucker a lot, creating a bristly tangle at the bottom of the trunk. Some species sucker much more readily than others: the field elms of England sucker vigorously, while wych elms do so much less.

The roots of the tree have many functions: they are the structural foundations, cleaving to the earth and digging down to counter the weight above, giving the tree stability and the ability to survive the pressure of wind without falling. The Dutch images show that elm tree root systems are not necessarily very deep but they will extend metres beyond the tree's canopy in a broad circle like the base of a central-legged table.

They also act as the water source, soaking up rain that falls or drilling down to find underground water sources. Elms may tolerate an occasional drought – some species, like the Siberian elm, more than others – but their lush foliage requires plenty of water, and something all the species in the family have in common is their preference for moist habitats. When they grow in dry areas they tend to be found around or in stream beds, and one of the main threats to the endangered members of the family is that climate change may desiccate their water sources. In arid parts of the USA, American elm has a tendency to invade cracked underground water pipes or sewers. On the other hand, they don't like to be completely waterlogged and will not grow in marshy conditions or places where rainwater doesn't drain away.

Roots are also food providers, either by directly ingesting minerals from the ground or by providing habitat for fungi that inhabit the soil in mycelial webs and swap minerals or other nutrients with the tree in exchange for sugars made by photosynthesis in the leaves. Elm trees will also help each other out by 'joining hands' underground: the roots of neighbouring trees naturally graft together, enabling all kinds of substances to flow between them. Our understanding of tree connectivity and communication is still very much in its infancy but the work of silviculturists like Suzanne Simard and Peter Wohlleben is helping us to see trees not as lone individuals but as members of communities, who help each other to thrive, responding to the needs of other trees, sending information about threats and opportunities. The underground network of roots between elms helps strong trees or those nearer resources to channel nutrients or water to hungry neighbours, but it also, unfortunately, can act as a conduit for diseases.

Over recent decades it has become widely understood that there is a 'wood wide web' of complex interconnections between trees in forests, thanks to their close relationships with fungi in the soil. Tree roots act as hosts for fungi in a symbiosis called mycorrhizae, giving them shelter and usually sugars or other food in exchange for minerals that the fungi extract from soil particles. The fungi exist as networks of tiny filaments, called mycelia, winding their way through the earth, sometimes across large areas, and we now recognise that these can act as chemical conduits between trees, enabling resources and information to be shared between them. Some of these fungi occasionally and seasonally produce fruiting bodies that we recognise as mushrooms and toadstools, but unlike trees like birch and oak, which host many of our favourite edible species like boletus and chanterelles, elms only have endomycorrhizal relationships with microscopic fungi that do not produce such fruits. Nevertheless, elms do sport a good range of handsome bracket fungi, and some edible species, notably *Pleurotus ostreatus* and *Pleurotus cornucopiae* (oyster mushrooms) and *Hirneola auricula* (jelly-like ear fungus), like to grow on them. Their deadwood is a favourite for many dramatic species, including *Rigidoporus ulmarius*, with its strange, hoof-shaped fruit, one of which, in Kew Garden, broke records by growing to a monstrous 316 kilograms at the base of a large, dying elm. Another is the gorgeously named Dryad's Saddle, *Polyporus squamosus*, which really does look like a woodland spirit could ride a branch away perched on its leathery, golden seat. Turkey tail (*Trametes versicolor*) might in some cases bear a relationship to its namesake, though it takes a stretch of the imagination, but it is nonetheless a lovely lacy thing valued for its medicinal properties in South-East Asia. These are just a few of the many hundreds of fungi known to associate with elms, and species new to science are still being found; two new species that grow on fallen elm leaves were recently identified by an amateur mycologist in Scotland, and in vast areas of the tropics and subtropics there has been nothing like the density of fungal exploration in the UK, so there are undoubtedly many other as yet unidentified species. Who knows how many fungi there are globally with close dependencies on elms or what discovering them might teach us about trees?

Unlike the intimate relationships between fungi and their host trees, lichens tend to use trees as just 'something to sit on', as lichenologist Brian Coppins points out in his important work on elm lichens, with the chapter on lichens in Max Coleman's *Wych Elm* book another relevant example.[5] However, they can be remarkably picky about which trees provide them with the right conditions for their perch. Some thrive in deep shade, some are greedy for light; some go for a smooth surface while others like it rough; they each have their own preferences for acidity and so on. One of the remarkable things about trees is the way in which their geometry enables them to manage the rain that falls on them: their branch and twig structures act as a complex network of gutters and drains so the moisture levels in different parts of trees of different species can vary widely even when the same rain falls on them. Back when I programmed mathematical models of trees, I wrote a paper about this, starting from the basic idea that a tree can either act like a big umbrella, in which the rainwater is mostly channelled outwards and away off the tree, or like a big funnel, in which the rain mostly ends up pouring down the trunk to the central roots. From the point of view of someone sheltering from rain under the tree, the effect is the same – as long as you don't lean against the trunk – but from the point of view of how much water reaches the roots, the two approaches are radically different. There are of course many complexities beyond this starting point, but you get the drift.

Lichens are bothered about such things and prefer certain locations on specific tree species because of the way that the tree is shaped to handle the rain and how much 'trickle-down' there will be. Elms have mostly rough bark with a dense, shading canopy and they have low acidity and high rainfall-retention capacity. This endears them to certain species of lichens, and although they aren't as rich in species as some broad-leaved trees, like oak, Brian Coppins says, 'This shortfall is often made up for by elm being the only host tree for some "specials".' Notably, in Scotland alone, 267 species of lichens have been found on elms and they are an essential habitat for thirty-seven species in the IUCN Red List of threatened species, seven of which are endangered. Two of the rarest priority species of lichens are *Gyalecta ulmi* (elm gyalecta), a pale, crusty thing with fruits that look like tiny, short-stemmed, white porcelain goblets full of red wine, and *Caloplaca luteoalba*, the orange-fruited elm-lichen, which genuinely looks as if someone has glued thousands of split red lentils onto the tree trunk. The first was only known to live on one tree, in Glen Lyon, which died, until it was discovered on an elm in Inverfarigaig in 2008, and it survives on a handful of calcareous rocky outcrops in nine other places. The second is known in only fifteen localities. The survival of species such as these hangs in the balance when elms are threatened.

Lichens matter for their own sakes of course, but they are also important because they are not single species. They are instead symbiotic relationships between fungi, which provide the structure, and algae which live in the structure. Algae have special abilities, such as being able to photosynthesise, so that fungi don't need to tap the tree for its sugars. Some lichens, including the lovely leafy *Lobaria* and *Peltigera* genera, are cyanolichens, with a blue-green alga as an alternative or sometimes a third species in the partnership, which can soak up nitrogen from the atmosphere. This is a powerful ability because nitrogen is the basic building block of all amino acids and thus of protein, but it tends to leach out of soils very easily and needs to be regularly replenished. When these species make their perches on elm trees, the lichens flake off and their nitrogen-rich flesh forms a nutrient-dense residue that is nourishing to the woodland ecosystem as a whole and the elm trees they have made their home.

Many of the lichens and other species that live on elms are not specific to them, as other trees provide a similar environment. Ash trees have comparable bark characteristics and grow in similar places and so there is a rich assemblage of elm-and-ash specialist lichens and mosses. In past decades there might have been some room for complacency about the loss of elms, given that many of these species were thriving on nearby ash trees, but with the onset and spread of ash dieback, another fungal disease that is spreading fast throughout Europe, coupled with elm's susceptibility to disease, this whole assemblage is now under threat. Without elms and ashes, many lichens and other epiphytic species will struggle to find a home.

Elms are also really helpful to invertebrates. Their early flowers are important sources of pollen for many flying insects such as honeybees at a time when little else is blossoming. In the USA it is estimated that more than 500 species of insects 'are thought to be intimately associated with elm by either breeding, feeding, ovipositing, or hibernating in elms'.[6] Some of these feed on elm exclusively, including the double-toothed prominent moth caterpillar which has evolved a perfect disguise, looking exactly like the toothed edge of an elm leaf. In the UK, elms host eighty species of invertebrates, including the delightfully named peppered light emerald moth and white-spotted pinion moth.

One priority species in the UK is the white-letter hairstreak butterfly, the caterpillar of which relies entirely on elm trees, and whose population was devastated when Dutch elm disease struck. Though also found in some elm-rich parts of continental Europe, it is very rare in the UK, being mostly restricted to England and Wales, but after having been unrecorded in Scotland for more than a hundred years, since 2017 it has been seen every year and is moving northwards. This butterfly is hard to spot as it lives mostly in the canopy of large elms. It has a distinctive white 'W' marking on the underside of its brown wings, with a gorgeous red flash at the wing tips, which have pretty, white-tipped black spurs. Its beauty isn't restricted to its wings, as it has delicate grey legs, a furry body and a black-and-white striped face like a tiny badger, and it is crowned with black-and-white stripey antennae with what look to be little flames at their points. It flies with a dazzling, swerving, spiralling flight. Colonies of hundreds of butterflies form around groups of elm trees, moving among them. They can sometimes be seen down at ground level if there are suitable sources of nectar. Although they mainly feed on honeydew that oozes from the trees, they also sometimes take nectar from nearby flowering shrubs or trees; they particularly like lime blossom. Some butterflies always rest with wings open and others with wings closed, and the hairstreaks all rest with their wings closed up, so their beautiful upper wings, much darker in colour, are rarely seen.

The hairstreak year begins early, just like the elm, with eggs hatching as the tree comes into flower. The eggs are each individually placed close to buds, or at the base of side shoots or on the underside of a twig where the new growth emerges from last year's wood. They are well camouflaged as they are small and brown, much like the buds, with a rounded shape reminiscent of a flying saucer. The larva, or caterpillar, initially feeds on the growing flower buds, and once flowers are fertilised it settles into eating young seeds – elms produce so many seeds in their clustering bunches, they can afford to lose a few. The growing larva then moves on to leaf buds, finally graduating to munching leaves when it is a bit bigger and can take on tougher fodder. By then it is a multi-segmented caterpillar with angled stripes that are supposed to make it look like a rolled-up leaf, but to me it looks like a tiny green accordion – it would be a tragedy to lose these lovely little squeeze-boxes. Once the larva is fully grown, by May or June, it pupates, creating a tough, dark-brown sleeping bag in which to undergo metamorphosis, one of nature's truly spectacular miracles. In late June to July the pupa splits open and the butterfly emerges to dance and feed, mate and lay eggs. Often two larvae of opposite sexes pupate close to each other – as if they form a lovers' bond as caterpillars, though they are unable to consummate their relationship until they have undergone their magical transformation and emerge as adult, flying beauties with 'W' for wonder etched onto their wings. Their main purpose in this brief, late-summer festival is to mate; I like to imagine they do this on the wing, though I am probably being fanciful. As autumn begins, after they have mated, the females carefully lay next year's eggs in promising nooks and crannies of the twigs, trying to make them blend in, disguised as buds. Just as the elm is profligate with its flowers and seeds, enabling it to be generous as a host to insects like the white-letter hairstreak, the butterflies must lay many eggs to survive the appetites of birds like blue tits, great tits, robins and wrens, which will have the winter to hunt them out. Once the eggs are laid, the short, sweet life of the hairstreak butterfly is over.

The early flowers may be important for insects, but it's the early seeds that follow which are a valuable food source for birds and squirrels. Perhaps most famously old elm trees, which often hollow out or develop holes, provide homes for all kinds of species. There is a delightful video from artist Robert Fuller, who films wildlife visiting a chunk of hollowed-out elm that he positioned up a tree in his garden to host owls.[7] As well as the most beautiful tawny owls that use it as a nest site, there are visits from kestrels, pine martin, squirrels and numerous small birds including treecreepers and blue tits, all foraging for insect life thriving in the hollow and decaying wood. Without elms, these many species have a harder time finding nest sites and miss out on a feast.

In North America, elms host one of the most beautiful birds, the red-headed woodpecker. This spectacular bird has a bright scarlet head, neck and bib on top of a white chest and splendid suit of black-and-white wing- and tail-feathers. It was formerly very widespread in east-central USA and southern Canada, so much so that the great ornithologist John James Audubon said of it in 1836, 'It is impossible to form any estimate of the number of these birds seen in the United States during the summer months; but this much I may safely assert, that a hundred have been shot upon a single cherry-tree in one day.' Sadly, the woodpecker is by no means so common these days, partly no doubt due to early birders shooting them by the hundred, but primarily due to loss of suitable habitat. It needs a hollow to nest in, and prefers snags or naturally hollowed trees, although the male will valiantly peck out soft wood if necessary and has been known to resort to hollowing out utility poles. But a naturally hollow elm tree makes life so much easier for the birds, and they were undoubtedly impacted by the huge losses of elms to Dutch elm disease in the 1960s and 70s. In the forty years from 1966, their population fell to just a third of previous numbers and they were listed as threatened on the IUCN Red List in 2004.[8] In recent years, although their population continues to decrease in some parts of the USA, the rate of decline has slowed due to careful management of their habitats, in particular leaving dead trees standing, and since 2018 they are now no longer considered under threat, although they remain threatened in Canada.

And let's not forget that elms are home to the most charismatic of forest species, once again because of their tendency to hollow out. Across the Northern Hemisphere, brown and black bears need winter homes because although strictly speaking they don't hibernate, they do fall into a state of deep torpor and need a safe and sheltered space to sleep through the cold weather. A mature elm tree, rotten in the centre, provides a perfect cavity for a bear to curl up in and snooze away until spring comes.

The Torbreck Elm

The nearest elm to where I live in Assynt, in the Scottish Highlands, is on the croft of one of my neighbours, in Torbreck. Crofts are small pieces of land, generally rented rather than owned. Conventionally a group of nearby crofts is called a township, and as well as the individual patches, or in-bye, there is a larger area of land, called the grazing, managed communally. Torbreck and the township I live in, Achmelvich, share a common grazing. Both townships are well-wooded, with a highly diverse mix of birch, hazel, willow, rowan, aspen, oak, holly and ash with a few cherry, juniper, hawthorn, blackthorn and alder found in odd corners. The underlying terrain is craggy gneiss – at about 3 billion years old it is one of the oldest rocks on the planet – and among bouldery scree above the road through Torbreck there are a few elms.

The most beautiful elm is on the west side of the road, growing on a precipitous slope. It is one of the tallest trees in the area and is as broad as it is high, with a grace quite remarkable for the steepness of its stance. In full leaf it dominates the glen, dwarfing all the birches, hazels and aspens that grow around it and seem to be clinging to the rock face, while the elm's great boughs sweep down with calm grandeur. Between the main trunk and the roadside fence, a smooth mound makes for a relatively easy clamber, and I guess this mound is actually made by a huge root bracing the tree against the rock. Once I have scrambled up the slope and established that even two people could not join hands while hugging the massive trunk, I find a place to sit round the far side of the tree on a rock. Either it tumbled down from the crag and broke its fall against the trunk, or the tree grew up beside it. Either way, a moss carpet has grown thickly over it and up the trunk, and as the top of the rock is almost level, it's extremely comfortable.

I lean back against the trunk and relax into the green light under the marquee-sized canopy. It's July and leaf cover is at its maximum. There is infrequent traffic on the road to Lochinver, but the cars are far down the slope and in the gaps between them all I hear is the trickle of the stream below, the hum and buzz of insects, the cheep of a chaffinch.

In the crack between the rock and the tree, herb robert grows among the moss, with its clubby little leaves and delicate pink flowers. I start to pay attention and acknowledge the diversity around me, the huge range of other plants being nurtured in the shelter of the elm: wood sorrel, dog violet, cow wheat, speedwell, goosegrass, chickweed, bracken and lady fern. On the upper edge of the canopy the elm's branches reach their closest to the ground, and around them, out of the shade, there is a ring of bright flowers: sunshine-yellow ragwort, St John's wort and nipplewort, purple-headed thistles and self-heal, a white froth of valerian and meadowsweet, deep pink fox gloves and a delicate lacework of grass seed heads. Above that a honeysuckle tumbles off a sheer section of the crag, a golden, sweet-scented curtain of blossom.

I can see where the previous crofter attempted to coppice some hazels, but the stumps haven't grown back – the pressure of grazing animals, deer and sheep, too great for new growth to survive. All the other trees grow at implausible angles out of cracks in the steepest rock sections, where teeth can't reach them. I notice that the elm foliage is bitten off to deer-height. Fortunately the vast bulk of the leaves are far out of the reach of herbivores. I wonder what protection from grazers enabled this tree to get established in the first place. It is probably old enough to have originated in the time when livestock were actively herded here, not just put out to fend for themselves and create whatever mischief they can get away with.

I look up through layer upon layer of tessellating green. The leaves vary hugely in size: most of the biggest are at the twig tips, opportunistically stretching out into pools of sunshine to feed.

A small, slater-like creature scuttles across my notebook. I try to take a photo of it with my phone but fail. It is a mystery animal. It is good to realise that the elm has familiars with which I am unfamiliar. A smaller version of the same beast scuttles even faster across my page – child of the first or some other, even smaller, species? A tinier, quicker mystery, encouraging me to keep looking closely. A beetle flies by, slowly – too big for an elm bark beetle, even the large one. The tiny flowers of chickweed and some golden-yellow pimpernel catch my eye. The tapestry of this undergrowth is so rich and luxuriant.

Above me, leaves are in constant motion, ruffled by the gentle breeze as if jostling for light. Some thin branches drooping into the space below the canopy sway and swing, bouncing with an irregular, syncopated jazz dance that is part of the orchestral complexity of the tree. The whole complex of movement appears to constitute a kind of language, as if I am seeing a visual, kinetic poem far more intricate than I can comprehend. If the tree feels each individual leaf tremor, like we feel drops of water on our skin, then what stimulation it must get from its tens of thousands of little leaf motions! The tree rustles as if in affirmation of this thought. Above us both, high on the crag, an aspen responds to the same breeze with hysterical fluttering, and a birch, even higher, up on the skyline, nods its graceful crown. It's quite likely that under the soil they are having even more intricate conversations.

The scent of honeysuckle and heather is heady and sweet, the light soft and warm. One of the cockerels crows from a croft up the road towards home and I begin to think about tea. Moisture starts to make itself felt through the seat of my trousers – the driest moss is rarely completely dry – but I feel deeply at ease here.

I am humbled and grateful to the tree for the comfortable seat, for the green light and sweet-scented air, for the rustling conversation I am too stupid to understand but not too stupid to ignore – perhaps when I have communed with more elms I will comprehend a little more. I am grateful for the sense of peace this tree gives me in its health, solidity and movement, for the oxygen it breathes out and for its conversion of carbon dioxide into sap and wood and roots and soil and cool shade to provide for the other animals and plants around me on a hot summer's afternoon. Most of all, I'm grateful it is so much bigger and older than I am.

Chapter Three

Death: Dutch Elm Disease

Although elms are big trees, some crucial things about them are tiny: their pollen is best viewed under a microscope, their seeds weigh only a fraction of a gramme and their biggest threat requires an imaginative exercise to envisage – it is the spore of a microfungus that is invisible except with modern scientific equipment, although its impact is vast and global. It has killed hundreds of millions of trees with what we call Dutch elm disease, with its telling acronym DED. Even the beetles of various species that spread the fungus are all less than a centimetre long and their eggs are just micrometres in size.

As we have seen, elm trees, like other trees, mostly live in harmony with and as generous host to other species, including insects and fungi, so this chapter will attempt to come to an understanding of why this particular species of fungus has been so devastating. It was originally named *Ceratocystis ulmi,* later renamed *Ophiostoma ulmi,* but the taxonomists decree that these days there is a more virulent and pervasive variant known officially as *Ophiostoma novo-ulmi*. Although it is deadly, exactly where, when and how many trees it kills is certainly not a given. After several years of the COVID-19 viral pandemic, most of us have learned how a virulent disease can spread, and how it might be contained or prevented in future. We also understand the crucial distinctions between a disease as a global phenomenon, the progress of which is measured by impacts on entire populations, versus the very different idea of a disease at the level of the individual victim. We're also savvy about the different susceptibilities to and experiences of an illness, the ways in which an individual can avoid or mitigate infection and the possible influence of this on the wider spread or impact. The same is true of Dutch elm disease.

Sickly trees were first noticed in several European countries, including the Netherlands, France, Belgium and Germany, just as the First World War was ending. Early research into the disease was carried out by a team of scientists at Utrecht University in the Netherlands: Dina Spierenburg, Marie Beatrice Schol-Schwarz and Christine Buisman. It seems remarkable to me that this team of three women plant pathologists carried out this groundbreaking research at a time when my alma mater University of Oxford didn't even allow women to be awarded undergraduate degrees. Their teamwork was fascinating. Dina Spierenburg carried out a detailed analysis of diseased elms, published in 1921, using photographic evidence to show that Dutch elm disease had actually been around in Europe since about 1900 and how it affected both the exterior forms of trees and their wood. Her colleague Bea Schol-Schwarz worked on the fungus causing the disease, isolating the spores of *Ophiostoma ulmi* in 1922 as part of her PhD research, which was led and supervised by the history-making Johanna Westerdijk, the first-ever female professor in the Netherlands – and one of the first female professors in the world – who raised the funding to enable this research to be carried out and nurtured her protégées. Westerdijk's contribution to the science of mycology is recognised in the world-leading Westerdijk Fungal Biodiversity Institute, which houses an immense collection of fungi and other microorganisms, including yeasts and bacteria that act as the international standard for identification of such organisms.

There was considerable debate in the Dutch and German scientific press about the initial results, so Johanna Westerdijk engaged another of her PhD students, Christine Buisman, to carry out a confirmatory study, published in 1927, that proved the fungus was indeed the cause of the disease. Buisman went on to spend a year at Harvard in Cambridge, Massachusetts, where she was able to prove in 1929 that the same fungus was present in North America. She then returned to the Netherlands and took up a position as the main researcher for the Netherlands Committee for Study and Control of the Elm Disease, becoming Europe's most important elm disease specialist. She set about trying to breed elms resistant to Dutch elm disease, developed a method for inoculating trees with the disease and bred and tested thousands of elm seedlings in the search for immunity from the disease. She died tragically after surgery aged just thirty-six. The first-ever fungus-resistant clone (her Clone Number 24) was named after her in 1937 by those who continued her work. The clone was widely planted, and although there are still many *Ulmus minor* 'Christine Buisman' trees to be found around Europe, their resistance to the original fungus has not always been sufficient to survive more virulent variants that have evolved. She is remembered by the Christine Buisman Foundation, which continues to support women who want to study to become scientists, particularly biologists.

The identification of the disease in the Netherlands is why it is dubbed 'Dutch' elm disease; however, it is believed to have originated in Asia because there are some species of elms there, including *Ulmus parvifolia* (Chinese elm) and *Ulmus pumila* (Siberian elm), that appear to have some immunity to the disease. This resistance is assumed to have evolved due to having the fungus present in their environment over a very long period, so that only those that did not succumb survived to become the ancestors of current trees.

The disease had probably been present in Europe for many decades, possibly even hundreds of years, but perhaps in a less virulent form or in such low density that it had little impact other than occasional local outbreaks, but the strain that was identified in the Netherlands in the 1920s began steadily spreading across Europe. It was particularly devastating in the Netherlands because almost all of the elm trees belonged to a single clone of *Ulmus × hollandica* (a cross between field and wych elm), which had been specially selected for its hardiness in salt winds and extensively planted for shelter around human habitation, but which proved very susceptible to the blight. Within the first couple of decades half of the country's trees were dead. It spread east to Austria and Romania and south to Italy, where again it was a major catastrophe because elms were used there for living supports of grapevines in vineyards. It was also grown for shade and shelter and its foliage lopped for summer animal fodder, so it had been a widespread feature in the landscape. Similarly in Britain, especially in England, elms had been grown as hedgerow trees for hundreds of years, and from when the disease was first noticed there in 1927 until the advent of the Second World War, it killed up to a fifth of all the elms, leaving gaps across the rural landscape. After this initial spread, however, the disease seemed to die down, and for at least thirty years from that spike in infections in the 1930s it was considered by British foresters to be no longer much of a problem, apart from occasional, locally contained occurrences – so much so that in 1958, forester H. L. Edlin, in his landmark book, *The Living Forest*, said of the disease, after acknowledging that it had 'spread alarmingly' until about 1930, that 'happily . . . its progress is now so spasmodic and irregular that it has ceased to be of much concern'.[1]

Meanwhile the disease moved west, reaching the USA, with cases reported in Ohio in 1928. The key question to ask here is: how does a fungus cross an ocean? And the answer is simple: global trade. Despite the fact that the USA has abundant forests of its own, elm logs were transported across the Atlantic because of a fashion for 'burled veneer'. This is made using very thin strips of elm wood from a burl, which is an abnormal growth or burr, where the wood becomes deformed and intensely knotted, resulting in some extraordinary patterns. The wood is cut into wafer-thin strips and used for the decorative outer layer of a piece of furniture. Creative cabinet-makers worked with repeats of the pattern, often playing with symmetry to create really gorgeous and unique items. Burls are sometimes caused when a tree is under stress, such as through insect infestation or disease, but elm produces them often even while healthy. The glut of elm wood from trees felled as a result of Dutch elm disease in Europe led to some very beautiful furniture, which became fashionable, fuelling the American market for European burled elm veneer. Hence, ironically, one of the nations richest in its own timber resources began importing infested elm logs from Europe, spreading the disease from East Coast ports along roads and railways. Over the next fifty years, it killed an estimated 100 million trees, causing what forester David Karnosky, in his seminal review of the impact of the disease in 1979, described as 'environmental havoc'.[2]

In the USA, there are several species of elms, explored in more detail in chapter eight, but the American elm is the tree that holds the iconic status, and its many millions of deaths are what made Dutch elm disease into a national tragedy. It grew wild in the forests, as a dominant species across much of the country, and had real importance for its woodland role, but for many Americans, it was most meaningful where it was deliberately cultivated. Elms were the preferred species to plant in new urban and rural developments, for structure and for shade, to enhance the overall design. The shape of an elm tree, with its tall, vase-like elegance, is a perfect foil to the straight lines of buildings, even while young. As it matures, it develops a dense but not overpowering canopy, through which dappled light filters and under which is thus a pool of shade that feels fresh and green and healthy. Two straight lines of elms form a tunnel with light at the end, often compared to the arched form of a cathedral roof. Any street lined with elms (and there were many – 'Elm Street' is still the seventh most common street name in the country) thus becomes a shady avenue, no matter how basic or boxy the houses or buildings behind the trees.

When elms catch Dutch elm disease the whole effect is ruined. First of all, the graceful shape becomes irregular as the canopy wilts off and dies, leaving bare branches. In Britain, this is referred to as the tree becoming 'stag-headed', as instead of a pleasing curve, there are jagged antlers sticking out of the top. Lines of dead or dying trees were often lopped to attempt to prevent spread of the disease by removing deadwood. The result was something akin to lines of tombstones – the cathedrals replaced by graveyards.

It happened incredibly rapidly: by 1950, Dutch elm disease was first noticed as having reached as far west as the state of Illinois. Nine years later, 95 per cent of all the state's elm trees were dead. The cities of Champaign and Urbana were home to 14,000 elm trees, which were reduced to a mere 40 living trees. In Minnesota it was even more extreme: Minneapolis and Saint Paul, the Twin Cities, lost 280,000 trees in the course of a single year, 1977. This is akin to urban people experiencing industrial logging – people who had effectively been used to living in a forest suddenly found themselves in a clearing.

The dramatic and visual impact of these losses led Americans to calculate the cost in various interesting ways. The loss of shade, for example, is not merely aesthetic. It means that houses are no longer sheltered from the sun, and in many parts of the USA this increased heat is a serious issue. It is no coincidence that installation of air-conditioning boomed following the elm decline, with corresponding financial and energy costs. Management of diseased trees also involves labour and transportation, felling deadwood, limbing to try to preserve trees, and preventative culling and replacement with other species.

The loss of elm trees from American urban environments was intensely studied, and the result is a much deeper understanding of the benefits of street trees and park-woodland in cities. The reduction of shade leads inevitably to increases in temperature at street level, but the environmental impacts go much further than this. Shade is visually beneficial as it reduces glare, which can cause traffic accidents and is psychologically wearing. Trees soak up a lot of water from the ground and transpire it from their leaves. This effectively turns rain at foundation level into humidity above our buildings, which has many benefits, helping to reduce aridity in dry areas and causing considerably less run-off of water into drainage systems and sewers. Trees also interact with wind in helpful ways, slowing it down, filtering it or absorbing its energy entirely and converting it into that rather delicious noise of fluttering leaves. This means that warm winds in summer don't dry us out so much, and cold winds in winter don't cool us down, leading to a much more equitable climate. While trees make pleasing sounds they also buffer and reduce noises we don't like – their impact on traffic noise can be significant. The impact of trees on air pollution is also significant, as they mop up particulates and nitrous oxides, and although by reducing wind they can cause air pollution to stagnate in local hotspots, overall they are beneficial.[3]

All of these social benefits can, by those who believe in such things, be given a financial value. One of the most incontrovertible financial values of trees is their effect on property prices: the aesthetic attraction of a mature tree, or trees, in a garden can bring about a significant hike in the price of a house. So the overall economic cost of Dutch elm disease goes far beyond the bill for management and replacement of the diseased trees themselves, huge though that can be. As well as lost real estate value, it includes heating and cooling of buildings, drainage, air and noise pollution and traffic accidents. The total costs of these since DED arrived in the USA were estimated by the end of the 1970s to be billions of dollars.[4]

In order to understand the more recent spread of Dutch elm disease and what can be done about it, it is necessary to understand more about what the fungus does to the tree. The fungus causing the disease usually makes its way into the tree helped by beetles that bore under the bark, to release sap for feeding or in order to create nursery spaces for their young. The fungus consists of very fine threads, or mycelia, which initially inhabit these nursery spaces but then spread into the xylem of the tree. The xylem is one of two systems of tubes within every tree; the other tubes are the phloem. These both lie between the bark and the older wood of the trunk and form the sapwood, the actively living part of the tree's structure, akin to a mammal's blood and lymph system. The xylem is columns of hardened cells that carry liquid up from the roots of the tree to all of the growing tips of its branches, supplying the vital water that is needed for all of life and in particular for photosynthesis, the near magical process by which leaves soak up carbon dioxide from the air, combine it with water and, catalysed by sunshine, create sugars. The water carried by the xylem also has nutrients dissolved into it, soaked up from the soil or provided by helpful fungal symbionts, which the tree uses for making and repairing cells. Amazingly, the xylem is able to lift water from underground to the canopy of a tree without using any energy at all, simply capillary action and the syphon effect. The phloem are softer tubes that transport sugars from leaves up and down the tree to wherever they are needed. Together these two layers of tubes form the annual ring pattern of growth inside the trunk.

It is the xylem that gets bunged up by the Dutch elm disease fungus because, unfortunately, the tree attempts to stop the fungus from spreading with plugs in the form of bladder-like extensions called tyloses, which grow from the cell walls in response to a toxin, cerato-ulmin, produced by the fungus.[5] The effect of these plugs is to dam up the xylem, thus preventing the necessary liquid from reaching the upper parts of the tree, causing leaves to wilt and wither and eventually the tree to die. It's as if the tree has an overactive immune response to the fungus. An initial infection in one branch of the tree can be spotted because only the xylem that connects that branch to the roots is congested and only that branch therefore wilts, rather like the way a blockage in a water pipe in one part of the building may only affect some rooms while the plumbing in others continues to work perfectly. An effective early treatment can therefore be to remove a wilting branch, but if the fungus is allowed to spread throughout the trunk, the tree will block up all of its xylem and, effectively, die of thirst. As the tree dies, so do all the other living things that depend upon it, most of which don't have much opportunity to move elsewhere, either because, like lichens or mosses, they are fixed where they grow, or because, like certain butterflies, their camouflage – or that of their caterpillars – means they need elms in particular to live on to avoid becoming easy prey.

The fungus does not spread from tree to tree alone; it has a little helper. There are actually eleven known species of beetle that potentially act as vectors, although usually one or two culprits common to a specific area. In my part of the world the main one is *Scolytus scolytus* (the large elm bark beetle) or its little cousin *Scolytus multistriatus*. As their name suggests, the beetles inhabit the bark of trees. The adult beetles live on sap, which flows just under the bark, and to find it they cut into vulnerable spots, often the junction between a three- or four-year old twig and its parent branch. When a beetle has succeeded in tapping into a sap flow, it will give off pheromones, powerful scent messages to attract other beetles to the site, so this sociable little critter is rarely found alone.

Under the bark is also the preferred breeding site of the beetles. Randy beetles pair off, with the female tunnelling along the grain between the sapwood and the bark to create a nursery channel. Meanwhile the male is 'preparing a nuptial chamber', as naturalist Gerard Wilkinson, in his 1978 book *Epitaph for the Elm*, so nicely put it. After they have mated in this mini boudoir, the female lays around seventy spherical white eggs on alternate sides of the channel she has created. The adults' job is done. I wonder what they do then, in the days between egg-laying and death? Presumably they live a carefree life, with no childcare concerns, feeding on sap, rubbing up with others who have similarly completed their genetic destinies, blissfully unaware of the fungal spores they are distributing as they go about their business.

When the eggs hatch, the grubs start eating the wood, burrowing at right angles to the main channel, munching their way along to create what is called a 'brood gallery', a tiny tunnel initially just half a millimetre in diameter, widening slowly as the larva grows. Each of its siblings does the same thing, so the single central egg gallery becomes surrounded by a fan-shape of tunnels, radiating out from the centre like an outsized dandelion flower. When the larvae are fully grown, having increased in size tenfold and created a tunnel about 10 centimetres long, they make a bigger burrow to pupate in, and once their miraculous metamorphosis is completed, on some warm day between May and October, they emerge from the bark to fly off and look for sap, mingle with other beetles and find their mate for the whole cycle to repeat itself.

The appetites of a few tiny beetle-grubs are easy for a big elm tree to cope with in normal circumstances, but the fungus finds the brood galleries, rich in larva-dung, to be its ideal habitat. Its filaments spread through these tunnels and it produces minute, hair-like stalks with miniature match-head spore cases called coremia, covered in thousands of sticky spores. The fungus doesn't adversely affect the grubs, but as the newly hatched beetles brush past the coremia on their outward journey from their nursery, they pick up the spores and transfer them to the trees they subsequently feed and breed on, delivering them neatly to their ideal new home as they seek out the sap of another tree.

The full life-cycle of the beetle, from eggs to adult beetles, takes about six weeks, so in a place with a warm summer, up to three generations of beetles can hatch out in a year. Once temperatures drop, any remaining adults die off, and brood-chambers full of eggs will lie dormant until the following spring.

As mentioned earlier, after the first wave of Dutch elm disease in Britain in the 1930s and 40s, it died down to levels that did not particularly worry foresters. Although its impact had been severe in relatively self-contained pockets through the first half of the twentieth century, people were pretty relaxed about it. But in the 1960s trees in southern England started to die in large numbers, initially in Gloucester and Essex. Then it spread, fast, killing 10 million trees in the next ten years.

The culprit was a new variant of the Dutch elm disease fungus. Again, the disease had crossed the Atlantic because of the international trade in wood. It was traced to Canadian rock elm logs that were being imported to Britain for use in boat building – as explored in chapter five, elm is particularly valued by shipwrights. Like COVID-19, the fungus develops modifications as it spreads, and the more innocuous strain of the fungus from thirty years prior, *Ophiostoma ulmi*, had mutated in North America. Its much more virulent cousin, now known as *Ophiostoma novo-ulmi*, appeared in Europe and stormed through the remaining elm populations. This form of the fungus had developed two characteristics that made it much more aggressive. Firstly, it grew and spread faster through xylem, so it killed whole trees much more quickly than the previous form, which might linger in one limb to be spotted and dealt with before it took out the whole tree. Secondly, the fungus had developed a new structure with lots of aerial mycelia, giving it a 'fluffy' appearance that made it more effective at spreading its spores.

It soon spread to the Netherlands, then France, Germany, Italy and on down into the Middle East. In several countries there had been major efforts since the first wave of the disease to breed resistant elms, but the result, for example in the Netherlands, was that there were millions of clones of one tree, all genetically identical, which proved susceptible to the new strain of the fungus and died in vast numbers.

This time Dutch elm disease was truly shocking in its impact. In England and Wales and into southern Scotland, 99 per cent of all elm trees died, and as the disease moved inexorably from county to county, woodlands were stripped to skeletons and hedgerows became lines of corpse trees. Rachel Blow, who worked on the efforts to control the disease in and around Edinburgh in the early 1980s, said, 'My memories of the losses of massive elms further south still pull at my heartstrings, seeing hedgerow giants laid across fields like evening shadows.'

This pattern was repeated across Europe as the disease worked its way east and north. For a while people believed that field elms – of which English elm is one clone – were more susceptible than other species, including wych elm, the species native to Scotland and with a huge northerly range across Scandinavia, the Baltic countries and into Russia. Indeed, even today you can hear people parrot the myth that wych elm is resistant to the disease. The reverse is true: it is highly vulnerable to the fungus. However, there were other factors that saved wych elm populations in the 1970s and 80s, most importantly the fact that they tended to grow in temperatures that were – and in some cases still are for now – too cold for the elm bark beetles to fly in. A combination of trade in infected logs and climate change has led to the disease sweeping steadily into wych elm territory, with the Baltic countries, like Estonia and Latvia, affected by the 1990s and the disease now reaching deep into Russia. In Norway elms are now endangered and in Sweden they are critically endangered.

The impact is such that even in otherwise dry scientific reports about the disease and its fungal and insect causes, it isn't uncommon to read about it 'ravaging' woodlands in a 'catastrophic' manner. When even mild-mannered foresters beat their chests, use strongly emotive language or wax poetical in grief, you know things must be bad. These trees have been iconic features of our landscapes, not just decoration but a key part of human habitats in many parts of the world, so when they die it can feel like our home is falling apart. When I started telling my friends and family that I was writing a book about elm trees, everyone seemed to have a story about a place that they love that was devastated by the loss of the elm trees – trees that we took for granted and so lost something unexpectedly precious, something we didn't know we needed until it was gone. There is a growing realisation that 'environmental grief' is more significant than we have previously understood: the loss of our natural homes wounds us deeply.

The loss also happens quickly. From the onset of an infection, revealed by some elm leaves on upper branches wilting and withering, it can be just a matter of weeks before whole branches are leafless and just a few more until the tree is totally defoliated. A tree that starts the summer looking healthy can easily be dead by autumn. You can go away on holiday and find your garden elm has died in your absence. This kind of rapid mortality adds a shock factor to the grieving process that makes it hard to sink in. My mother appeared fine until she felt sick during Christmas dinner at my sister's house. A doctor came to see her and thought she should be admitted to hospital, where she was diagnosed with advanced and untreatable cancer. She was dead three weeks later. There was just time to say goodbye, but not to take in that she wasn't going to be around anymore. Afterwards, her sudden absence left a gaping hole. Something similar happens when Dutch elm disease sweeps through a locality, taking out the lovely shade and shelter, plus the comfort and familiarity of trees that we expect to be there forever.

The raw numbers of tree losses are stark and the visual impact on landscapes shocking, but the knock-on impact on other species in the ecosystem most concerns many ecologists and conservationists. A detailed study of the impact on birds in English farmland was carried out in the 1980s by British ornithologist Patrick Osborne, who concluded that 'elm death does not seem a sufficient reason for vacating an area for many bird species', so the loss of the elms is not an immediate death sentence, not least because of the short-term opportunities to feast on the elm bark beetles and other insects that flourish in the bark of the dead trees and the plentiful nesting sites that appear in the rotting timber.[6] However, what we do about the death of elm trees is crucial. Patrick Osborne said, 'Elm felling is a wholly different matter. Besides herb and shrub upgrowth, there are no benefits from elm felling; all nest sites and food sources are removed, and habitat structure is dramatically altered.' The evidence suggests that sanitary felling and removal of dead elms may limit spread of the disease but at a cost to local biodiversity.

This finding is corroborated by a more recent piece of Spanish research into ants, carried out by Soledad Carpintero and Joaquin Reyes-López from the University of Córdoba.[7] They compared the fate of ant colonies living in and around riverine elm woods between 2007 and 2016. Dutch elm disease killed many of the elms from 2010, so they were able to watch as the size and species of ant populations changed due to the loss of the primary canopy in their woodland by the Bejarano and Molino streams and Guadiato River and a control meadow patch in Dehesa in Córdoba. Over the decade they identified more than 20,000 ants of 51 species. As with the birds, the elms' death did not initially alter the number of species of ants but, crucially, what was done with their remains had a huge impact on which ant species could continue to live there. Where the dead elms were removed, involving heavy machinery and significant landscape impact, common and generalist ant species increased but many of the 'interesting' and less common ant species like *Temnothorax bejaraniensis*, which are litter and twig specialists, were lost. Carpintero and Reyes-López reported, 'The loss of these species and in general the changes in ant assemblages in Bejarano after logging and harvesting dead trees were of such magnitude that its ant assemblage became more similar to the one of the meadows (Dehesa) than to the ones of the rest of the riparian areas.' In other words, when the dead trees were removed, only the ants that could live in open ground survived. Like Osborne, they found this impact is 'stronger after the felling of dead trees than after trees' death . . . because following the logging, the habitat structure was drastically altered and all nest sites and food resources were removed'. The ants on the woodland floor would clearly rather we left the elms standing even after they've died.

The twentieth century was not the first time elms suffered a huge wipe-out. Around 5,000 years ago, elm trees were widespread right across Europe, but then, suddenly, they were gone. The abrupt death of what must have been countless millions of elms across the continent is known as the mid-Holocene European elm decline and is, according to archaeologists Sarah Clark and Kevin Edwards, 'the most widely discussed phenomenon in post-glacial vegetation history'. The Holocene is the name given to the geological era following the most recent ice age, from roughly 12,000 years ago up to the current period, generally known as the Anthropocene, named for the fact that human impacts on the planet will leave a geological 'signature' at least as significant as any previous geological upheaval. During the early Holocene, there was a gradual movement of species, including humans, across the land as ice retreated and the climate warmed. Exactly what species were present is known thanks to painstaking analysis by scientists who uncover their tracks and traces in sediments and deposits in bodies of water and peatlands.

Palaeobotanists and specifically palynologists study the plants of the past, digging up samples of ancient dust scattered into places where it has been preserved from rot by low oxygen levels, then peering down microscopes at that dust, seeking out, identifying and counting the grains of very, very old pollen that lurk among it. It seems vanishingly unlikely that something as ephemeral as pollen could survive in a peat bog for several thousand years, but survive it does. Under a microscope a grain of elm pollen looks like a little round pebble, pitted and layered at its edge. It's a mere 25 micrometres across.

If you drive a pipe down into peaty ground or other land that has lain undisturbed for thousands of years and been wet enough to prevent aerobic decomposition, then pull up the pipe, inside it is a column of soil. The oldest material is at the bottom of the core, and using carbon dating it is possible to determine the age of each layer. Each slice of the core is thus a snapshot of a few years. Because pollen is wind-distributed, it gives a picture of which plants were growing nearby. Palaeobotanists can thus chart how, after the ice retreated, hardy pioneer trees like birch, willow and pine colonised the ground first, and later trees like oak, ash, elm and hazel joined the woodland mix. They can show that certain places were dominated by wetland species like alder and willow and then as land dried out less moisture-tolerant species like pine moved in. The fossil record reveals the human impact from grazing, burning and, later, cultivation of arable crops, and how this activity changed the species mix. These shifts are usually subtle and gradual, but the elm decline in the mid-Holocene wasn't subtle at all. Over the course of a very few years, elm pollen levels collapsed dramatically. A study from Diss Mere in eastern England demonstrated that 5,000 years ago there was a sudden loss of elms, with half of them vanishing in just five years. Elm's contribution to the pollen during this time plummeted from 7 per cent to less than 1 per cent in less than a human lifespan.[8] Laura Flynn and Fraser Mitchell's research, published in 2019, compares the pollen record from the ancient elm decline with the contemporary Dutch elm disease passing through modern landscapes.[9] The precipitous drop-off in pollen levels is statistically similar enough to suggest a disease comparable to the present elm blight, as opposed to alternative theories that human impact or climate change are to blame, both of which no doubt contributed to elm losses but alone would bring about a much more gradual decline.

Palaeoentomologists, who study the insects of the past, have even discovered the remains of elm bark beetles from the appropriate prehistoric moment. The presence of *Scolytus scolytus* in deposits from precisely the time of the elm decline, and the lack of any of these beetles in later deposits is, as Sarah Clark and Kevin Edwards say, 'clearly tantalising', although it 'neither proves nor disproves the role of the elm bark beetle in a disease explanation for the classic elm decline'.[10] But the weight of evidence, painstakingly gathered from across Europe, suggests an earlier wave of beetle-carried disease that ravaged elms in just the same way that Dutch elm disease has done in our lifetimes. There is, as yet, no trace of the disease-causing fungus itself, if a fungus it was, though it is tempting to assume that it was not the current culprit *Ophiostoma ulmi*, or its modern *novo-ulmi* variant, because if it had been, the elms that survived the disease would presumably have had and passed on resistance to it.

What can we learn from this historic wave of elm disease? It illustrates the tendency of elms to succumb en masse and hints at an earlier wave of beetle-transmitted disease, and thus implies that afflictions like Dutch elm disease are not a one-off. It also demonstrates that despite catastrophic decline, elm species as a whole need not necessarily be wiped out – although as we've seen from the fossil record, there were plenty of now-extinct elms in times past.

Max Coleman, Scotland's leading elm expert and author of *Wych Elm*, the definitive book about our native elm, refers to the 'symbiosis' between the beetle and the fungus. It's obvious that the beetle's activity benefits the fungus, as a highly effective vector of the spores from tree to tree. But in a symbiosis, the fungus should be benefiting the beetle too, and it is counter-intuitive that spreading a disease that destroys the tree you are dependent on could be in any way beneficial from the beetle's perspective. So I asked Max how the fungus benefits the beetle, and he said, 'The beetles love recently dead wood. It's easier for them to burrow under the bark and for the larvae to feed. Spreading the fungus creates a bounty for the beetles, at least in the short term.' In the longer term, in wiping out whole populations of the trees, the beetles look about as short-sighted in their environmental management as humans are. Max talks about the fungus, beetle and tree being involved in a complex three-partner evolutionary dance – each time the fungus makes a move, improving its way of surviving and spreading, the beetle and tree will need to change to keep in step. This takes time, but it gives me a glimmer of hope that even if the disease tears through the woods I love, it will leave some of the mighty elms standing to live on into the future. In the great ceilidh dance that is life on Earth, we need to ensure that we aren't that person who goes blundering across the floor, upsetting the rhythm so much that chaos results.

The Preston Twins

There is nothing quite like a huge fat tree, I think, for reminding us that our lifetimes are fleeting. A tree too big to hug on your own is a wonderful thing, and when a tree is so big it needs a team effort to reach all the way around it, it is special indeed. One such tree, with a girth of 6.5 metres, can be found in Preston Park in central Brighton, on the south coast of England. Along with a weeping elm and some other rarer elm species, it grows in the Coronation Garden, where it stands out from all the others due to its magnificent size. Back in 1613, it was planted as part of a row of English elms (*Ulmus minor* 'Atinia') put in as a hedgerow to shelter a manor house, and in the ensuing 400 years it has grown into a champion. For a long time it was one of a pair of equally chunky elms, and was the marginally shorter of the two trees known as the Preston Twins.

I visit it in early October and it is still full of leaf. It's clear that it has been pruned, quite severely, with some large limbs taken out completely for stability, but it has continued growing undaunted. Its shade is dense and cool. A busy road runs nearby, but round the park side of the trunk, underneath its canopy, the rustle of its leaves and its dense wood shield out the sound of traffic. I could happily spend ages lingering under here but I have come a long way to spend limited time with one of the country's key elm experts, Alister Peters, so I can't indulge for long in communing with this gentle old tree. Moreover, there is a far stranger tree waiting.

The sibling of this mighty tree once held the honour of being the fattest English elm at more than 7 metres in girth. It held a magnificent crown of leaves – and was a real giant. Then, in 2017, it suffered storm damage and lost a limb and a part of its trunk. The following year it became infected with Dutch elm disease. A trench was dug to separate its roots from its twin's to try to prevent the disease from being transmitted underground, and some drastic pruning went on, but to no avail. In 2019 the decision was taken to fell it.

What happened next is absolutely extraordinary. It has been turned into a work of art, a sculpture that retains as much of the trunk of the tree as could be rescued and honours it with a sheen of gold.

A green metalwork fence prevents a close encounter, but as I'm gazing at it longingly through this barrier, Alister produces a set of keys from his pocket, one of which unlocks a gate in the fence. Unlike its twin, this dead tree has a fur of undergrowth, which also disguises the fact that at ground level it's no longer attached to its roots. All that remains of the tree, in fact, is a hollowed-out shell.

One side of the trunk shimmers black as if recently charred, and it looks like someone has poured molten gold from the sky into its upper reaches. Around the back, there is a latticework of black metal across a hole in the trunk, revealing the inner surface, gleaming like a dragon's hoard. It is as if the tree's golden autumn leaf colour has blazed so fiercely it has become a magical kind of conflagration. This inner space is as big as a stairwell that should spiral up to a turret in some magical castle.

When it was felled, Alister tells me, it turned out to be completely hollow inside, just like other ancient elms, which live on with hardly any of their original heartwood. The strength of such a tree, its ability to bear its canopy of leaves, must rely entirely on the living sapwood and bark around the outer circumference of that mighty trunk. The felled tree, with little wood to give it a robust structure, was extremely fragile, and its sheer scale made it a huge challenge to manoeuvre. Nonetheless, it was taken out in one piece and put into storage, while someone was sought to do something worthwhile with it and fill what was widely felt to be a 'gaping hole' beside its bereaved sibling.

Eventually it was given to Elpida Hadzi-Vasileva, an artist, who created 'a new sculptural celebration' of the ancient elm.[11] Working on the tree over the course of several years, including a year in the Secret Garden in Kemptown, east Brighton, she used an ancient Japanese process called *yakisugi*, involving debarking and charring dead trees to preserve the wood. She then gilded a large portion of the interior. The sculpture appears like a huge example of *kintsugi*, the tradition of mending broken pottery with gold-coloured tree sap, so that the cracks and joins become a distinctive part of the whole. To treat a dead tree in this way also seems to embody that wonderful Japanese concept of *wabi-sabi*, which celebrates imperfection and acknowledges the transience of everything. Writer Richard Powell, in his book on the concept, summarises it as follows: '*Wabi-sabi* nurtures all that is authentic by acknowledging three simple realities:

nothing lasts, nothing is finished, and nothing is perfect.' Elpida Hadzi-Vasileva has shown us how we can relish the tree's death, decay and scars.

She oiled the outside of the tree with black oil to highlight the patina of the wood. She consolidated and strengthened the inside and then painted it with gilt paint, so that when the sun hits, it gleams and shines, revealing the extraordinary textures and shapes in the timber.

The project was precarious and immensely difficult, both logistically and financially, involving three moves of the vulnerable tree in various stages of semi-preservation. But her commitment to the elm and its transformation into a gilded artwork meant that eventually it was returned to Preston Park, alongside its living twin. In the process it has become much more than simply a modified tree. It stands as a lasting memorial to all the elms that have died of Dutch elm disease. 'The trees may be lost,' Elpida says, 'but they need to live in our memory, knowledge and experience. This project is a final opportunity to hold and celebrate this disappearing past.'

In addition to the main sculpture, some small pieces of bark were also preserved and gilded with 23.5 carat gold leaf, attached to small elm plinths, turning what might otherwise be considered bits of waste, sloughed off and discardable, into collectable art treasures. What greater honouring could an old tree be given?

Chapter Four

Life: Healing Uses

In modern society, when our health suffers, we mostly seek help from the medical world, visiting clinics, surgeries or hospitals, where doctors prescribe treatments that are often pharmaceutical. We take their prescriptions to counters where white-coated professionals exchange them for foil blister packs of tablets or bottles of medicines. Many people, especially older ones with complex health issues, may be consuming a rattling handful of such drugs on a daily basis for years.

It's trite to say that it hasn't always been like this, and I'd be an idiot if I wasn't grateful for modern medicine and its ability to diagnose, treat, cure or ward off a vast range of illnesses and injuries. But in the process of medical developments in many Western countries, knowledge of the healing powers of our companion species in the natural world has been given short shrift. In Britain and other European countries there used to be people in every community, often women, who were expert in herbal and natural lore, and could provide relief with a well-chosen concoction of plants. Unfortunately, those skills were demonised as their roles as spell-casters and believers in natural spirits set them at odds with leaders of organised religious institutions. The systematic persecution of witches in the Middle Ages caused a catastrophic loss of herbal knowledge from much of Western Europe and a cultural suspicion towards any kind of medicine that doesn't come from a lab.

Fortunately, some refused to forget what the natural world could provide, and a few early medical professionals gathered herbal knowledge and wrote it down for posterity. One key figure in Great Britain was Nicholas Culpeper, born in Surrey in 1616, a physician who used a mix of knowledge from astrology and botany, running a pharmacy in Spitalfields, London, offering medical help for free. He gathered herbal medicinal knowledge into a compendium, *Culpeper's Complete Herbal*, published in 1652, in English, as a practical demonstration of his philosophy that this knowledge should be made widely available to anyone who could benefit from it – a radical departure from the elite medical world, which circulated its knowledge in private journals, mostly in Latin. He died, as so many did, of tuberculosis, aged only thirty-seven, but in his short life he catalogued the healing properties of thousands of plants, including elm.

'It is a cold and Saturnine plant,' he says of elm, the bruised leaves of which 'heal green wounds, being bound thereon with its own bark'. Green wounds are those that have gone septic or become infected. Culpeper's entry for elm is shorter and simpler than for many plants but it is emphatic that elm is a powerful healer of skin conditions, treating 'scurf and leprosy very effectually'. It is also 'very effectual to cleanse the skin' he insists, repeating that a liquid extract of the leaves 'is a singular and sovereign balm for green wounds'. His instructions for conjuring this extract are somewhat mind-boggling, involving burying a glass vessel in dung for twenty-five days, but pharmacy labs weren't quite what we've come to expect these days. Culpeper also asserts that a decoction of leaves, bark or root will heal broken bones. His final recommendation is that a fomentation (poultice) of the bark of the root can treat tumours and address 'the shrinking of the sinews', presumably meaning conditions such as Dupuytren's contracture, which can cause fingers to contract towards the palms.

For Culpeper, all plants were linked to one of the planets, and this astrological connection was key to understanding herbal treatments. Elm was connected with the planet Saturn, which meant that its characteristics were 'knobby, knotty or hard' and it was a plant of coldness and darkness, strongly linked with melancholy and old age. Most modern herbal medicine practitioners don't make any such connections with astrology, though there are exceptions, but accepting the astringent quality of elm leaves and their efficacy in treating wounds doesn't require any belief in the intervention of a distant planet or its associated spirits or deities. Most books of tree lore make reference to the use of elm leaves for dressing wounds, though many simply repeat Culpeper's recommendations, but occasionally there are additional instructions, such as bruising of the leaves or pickling them in vinegar.

One of the intriguing peculiarities of Culpeper's note on elm is a reference to the healing qualities of a liquid that can be found in 'bladders' on the leaves, and it's possible this knowledge has indeed been lost. Jacqueline Memory Paterson's *Tree Wisdom* states that blisters on English elm leaves are 'caused by a leaf-louse', of which I can find no trace. The Royal Horticultural Society blames the elm pimple gall mite (*Aceria campestricola*) for 'small whitish-green hard raised structures on the upper leaf surface', which don't seem like promising sources of a medicinal juice at all. There are other fascinating galls found on elm, including peculiar green protrusions called elm finger galls (*Eriophyes ulmi*) and very striking red and yellow growths that look rather like birch catkins or sometimes like a rooster's comb, hence their name of elm cockscomb galls, caused by aphids (*Colopha ulmicola* and *Colopha graminis*), but it's hard to imagine how anyone could extract any kind of liquid from them.[1] Much more likely is a yellowish bladder-like growth formed in midsummer by an aphid, sometimes called the elm balloon-gall aphid (*Eriosoma lanuginosum*).[2] This growth starts off looking rather like a green walnut but can develop to be up to 8 centimetres long and nearly as wide, sometimes swelling evenly into something like a ripe pear but more usually growing in a lumpy, irregular manner. Another possibility is a round green growth caused by the elm sack gall aphid (*Tetraneura ulmi*). I can find no evidence that anyone these days is extracting healing balm from such lumps and bumps on elms, but I'd be delighted to meet them if such a pharmacist exists.

Much less obscure than arcane preparations from protrusions on diseased leaves is the medicinal use of elm bark. According to a nineteenth-century account of the botany of the Scottish borders, one of the local names of elm is 'chewbark' and its inner bark is described as having 'a certain pleasant clamminess'.[3] Almost invariably, herbal texts that include elm make reference to the beneficial use of its bark, primarily for external use on skin conditions, including wounds and burns, or applied as a poultice for fractured bones. One online compendium of herbal medicine traditions suggests that bathing in elm bark and alum is effective for reducing heavy menstrual bleeding.[4] It also includes drinking a decoction of elm bark for purging phlegm and treating diarrhoea. There seems no end to the benefits of a dose of elm bark – apparently it even tackles baldness!

Most of the herbal medicine texts don't specify which elm tree species they are referring to. Sometimes constraints can be deduced from geographical location: it's unlikely that Culpeper came across a wide variety of elms in London in the seventeenth century, and it's likely he meant the English elm (*Ulmus minor* 'Atinia'), but even so, the recipes he was gathering could in theory have applied to the several elms found in Britain. Herbals that draw on the indigenous traditions of other regions, notably North America, can cause greater confusion, possibly referring to a much wider range of elms.

However, one North American species frequently mentioned for its medicinal properties, particularly for its bark, is slippery elm (*Ulmus rubra*, sometimes called *Ulmus fulva*). The tree is identifiable by chewing a twig to check it has a sticky texture and sweet taste, and its fragrant inner bark is particularly valuable. Slippery elm bark is readily available in powdered form from health food suppliers. It is high in calcium and has been used as a milk substitute for babies; presumably, its easeful properties are beneficial to all.

According to Ellen Evert Hopman, a Massachusetts-based herbalist and druid, slippery elm 'eases insomnia and is soothing to the stomach and bowels'.[5] She recommends adding blended sesame seeds or almonds to the milk, or making a pudding with it by adding egg, honey and flavourings such as cinnamon, nutmeg or lemon rind. A long list of digestive ailments is apparently eased by regular consumption of such dishes. There are also some eyewatering recommendations for the opposite end of the canal, with slippery elm forming a key ingredient in various herbal enema recipes, the details of which I shall not dwell on. The benefits of slippery elm bark are not limited to the guts, however, as it is also recommended in cough medicines, for bronchitis and lung disease, and even against typhoid fever, if mixed with honey and cayenne pepper.

Like elm leaves, slippery elm bark has a long heritage as a cure for skin ailments, and it was a key component of surgeons' medicine chests during the American War of Independence, being used as a poultice to treat wounds, including from gunshots. It is said to be useful for tackling inflammations and infected flesh, being sometimes effective even against gangrene. The key to its effectiveness seems to be its mucilaginous or slimy quality, which draws out infection from wounds or sores or mucus membranes in the gut or respiratory system. Along with other elms, including the Himalayan and Siberian elms, it is being seriously explored in the fight against antibiotic resistance because it is effective against certain bacterial and viral infections, at least for some people.

In the interests of research, I tried it. The packet gave instructions for a hot drink, predominantly milk. Now, sweet milky drinks are not really my cup of tea – which I take black, weak, without sugar – but I dutifully boiled up my soya and mixed in the floury powder, which turned it vaguely grey. It tasted not unpleasant and faintly woody. I've licked and chewed a fair bit of bark in my time as a crazy, tree-hugging poet, and this tasted quite similar to willow, though less bitter. The drink was greatly improved by the addition of a couple of heaped teaspoonfuls of chocolate powder! It left me with a warm glow in my belly and no particular desire to repeat the experiment. I feared that the rest of the packet was destined to lurk in the back of my food cupboard until it went mouldy, but it occurred to me to try it in bread. Several loaves later, I have emptied the box and have nothing but good things to say about it. I recommend substituting it for about 20 per cent of the flour.

Unlike the leaves and bark, there are relatively few medicinal references to the use of elm flowers, although one notable exception is in the Bach flower remedies. These are thirty-eight flower essences that were selected in the 1920s by Edward Bach, a doctor and early exponent of homeopathy, who believed that unsettled emotions were the root cause of all physical ailments and ill health. 'Behind all disease lie our fears, our anxieties, our greed, our likes and dislikes,' he said. 'Let us seek these out and heal them, and with the healing of them will go the disease from which we suffer.' He recommended tinctures of particular flowers to address a wide range of emotional states seen as root causes of disease. 'Take no notice of the disease, think only of the outlook on life of the one in distress,' Bach urged in the introduction to the short tract he wrote about his system.[6] In this system, elm is one of a group of plants, along with larch, pine, sweet chestnut, star of Bethlehem, willow, oak and crab apple, that is used as treatments 'for despondency and despair'. All of these woody remedies are prepared using a boiling method. The flowers should be gathered into and floated on fresh spring water in a glass bowl without touching them, either by holding the vessel under the blossoms and snipping them directly onto the liquid, or by covering your hand with a leaf so your skin doesn't make contact with the healing blooms. Once the surface of the water is covered, the water and flowers are boiled for half an hour, then the strained fluid is poured into bottles, half-filling them and topping up with brandy as a preservative. These bottles are 'mother tincture' from which 'stock' bottles are made up using two drops of tincture to 30 millilitres of brandy, delivering a vanishingly small concentration of the essence of the flowers, which is a characteristic of the homeopathic method.

The thirty-eight remedies include various poetical connections between plants and emotions – notoriously, impatiens is used to address impatience – and elm flowers are recommended to address 'overwhelm', particularly in capable people who have taken on too much. Bach's definition of people who may be helped by the elm homeopathic remedy is as follows: 'Those who are doing good work, are following the calling of their life and who hope to do something of importance, and this often for the benefit of humanity. At times there may be periods of depression when they feel that the task they have undertaken is too difficult, and not within the power of a human being.'[7] I can recognise myself in this description! A dose of two to four drops of this elm flower remedy, four times a day, under the tongue or in a drink, is supposed to bring about a lifting of the sense of exhaustion and dejection that comes from having too many responsibilities and being unable to fulfil them all.

As I'm willing to try most things once, I put myself on a course of elm flower remedy. It coincided with a particularly busy patch of editing this book and I could happily persuade myself that the microdoses of elm tincture were indeed helpful. I managed to be highly focused and, although I was tetchy and irritable, I wasn't dejected. I'm a great believer in the placebo effect!

I'm not advocating the Bach system, nor do I want to cast aspersions on it or on any other body of herbal medicinal lore. There is much in the world that we do not understand, and many apparent 'old wives' tales' or superstitions end up having roots in science. My favourite of these is the old taboo about bringing hawthorn or May blossom into a house, based on colourful stories in several cultures about a death goddess or crone spirit who will murder babies if this is done. It turns out that hawthorn blossom gives off an aromatic gas, trimethylamine, which is also produced by rotting corpses and is attractive to small black flies. Should a dead animal be decaying nearby, flies that have been feeding on it may subsequently flit in through an open window, drawn in by hawthorn blossom, and thereby transfer disease to any unsuspecting infant. The old wives are vindicated.

Although elm leaves, bark and flowers are good for specific aspects of our health, the seeds do not appear to bring any medical benefits at all. There are recipes that use them, such as one in Fi Martynoga's *A Handbook of Scotland's Wild Harvests*, a fascinating compendium invaluable for Scottish foragers, drawing together knowledge of edible native species from a range of experts.[8] Ian Edwards contributes a 'Chinese-inspired' suggestion of dipping wych elm seeds in flour and steaming them like dim sum, recommending that they are harvested young before their samara wings become too papery. They are, he assures us, a 'delicious nutty snack', if a bit fiddly to prepare, and are no doubt nourishing. He also makes reference to the edible inner bark, but refers to it as a famine food once eaten by those with no other choice, noting the irony of the high price tag of American slippery elm powder.

There's a video on 'Sister Xia', the YouTube channel of Xia Jie from northern Shaanxi province in China, in which she makes 'cutlets' out of what are referred to as elm leaves but which are clearly the immature seeds, collected in spring while the cherry trees are in full blossom.[9] She mixes washed seeds, with their papery wings still green and tender, with carrots, eggs, flour and spices to create sticky balls, which she steams for fifteen minutes then serves with a garlic, chilli and onion sauce. By the looks of the grins on her family's faces, they're delicious. The website A Food Forest in Your Garden raves about elm seeds having a 'succulent oiliness' and recommends using them in salads or in a green-leaf curry sauce.[10] I have to confess that my own efforts at munching on elm seeds doesn't lead me to recommend them – suffice to say, I found myself picking inedible papery skins out of my teeth! I suspect that timing of the harvest matters a great deal; perhaps I left it too late in the year and not all species are equal, with a lot of variation in seeds. Nonetheless, elm seeds are beautiful things, spinning in the wind, lifting our spirits and thereby nurturing our health and wellbeing as much as anything in the herbalist's pharmacopoeia.

There's a terrible irony that a species that does so much to heal us can itself be so badly beset by disease. Fortunately, it turns out there is plenty that can be done to help an elm tree withstand infection by Dutch elm disease, and arboriculturists and tree-lovers have been nursing them successfully, finding ways to protect and treat them and literally saving their lives. Early signs of the disease can be spotted by looking for dieback in upper branches in summer. Treatment by removal of dead or dying limbs can do a lot to check the spread, but it needs to be done carefully.

Infected elms that are felled do not necessarily die completely. Elms, like many deciduous trees, can easily withstand the loss of limbs and even in some cases the entire trunk, and will send out new shoots from the base of the tree. This is a characteristic that is exploited by the forestry practice of coppicing, when a tree is cut to its base and then allowed to resprout from its roots, or pollarding, when the trunk is cut above the height of grazing animals that might eat shoots at ground level. The result will never be a tree of the same elegant form as one with its primary trunk, but as long as the fungus hasn't reached the roots, the tree will live on, and from the perspective of many small animals and other life forms that depend on it, a multi-stemmed coppiced or bush-headed pollarded elm will still be a fine home. So there is hope here.

Some places have been particularly effective in limiting the spread of Dutch elm disease, and Brighton and Hove, where the Preston Twins hail from, is one of the very best examples. These days it is home to the National Elm Collection, with more than a hundred different varieties across many of the temperate species. The region began protecting elms in the 1970s, long before it achieved this status, partly due to the vision of one maverick and brave man, Ray Strong, who saw how much impact the disease would have on the town and partly due to a groundswell of public desire from ordinary people in the area to protect their trees. Brighton and Hove are adjacent towns, governed as a city, on a very exposed stretch of the south coast of England, where the underlying rock is chalk, and elm is therefore one of very few species of trees that will tolerate its harsh conditions. Alister Peters is employed by the local council to devote his summers to spotting and dealing with Dutch elm disease, and when I visited him to meet the Preston Twins, he took me up to a vantage point over the city. It was a beautiful October day, but as a sailor it was easy for me to see how the area is wide open to any weather tearing up the English Channel. He pointed out shelter belts of woodland and explained how the Victorians had planted elms along the streets to ameliorate the salt winds, then told me the story of how they had not all succumbed to fungal blight. At the time, the official national position was that the disease should just be allowed to run its course, and nothing could be done to deal with it. Alister said, 'Ray quickly worked out that if we lose the elms, it would be disastrous, because not only do you lose the elm, but you'd also lose a lot of protection which allows other plants to survive.' He firmly believed they needed to stop it spreading, and ganged up with a local teacher to form a pressure group called Save the Elm, which started taking direct action whenever they spotted diseased trees. Alister told me, 'They were cutting down street trees, park trees, trees in private gardens, the whole lot, just ordinary people with a scattering of some of the guys in the council tree gang. It was a people's uprising. They embarrassed the local authority into getting involved. And the Forestry Commission was saying this was never going to work. But it did.'

Initially there was a lot of experimentation to find out how the beetles and fungus behaved and how to most effectively control them. As Alister drove me through the town, pointing out various species of elms along the streets, he talked through the lessons they learned. A key one was that there needed to be a consistent approach across the whole area, and fortunately they persuaded the local authority that they needed to tackle all elms, whether on public or private land. 'Ray painted this picture of disaster, and he also said, "You can't just treat the council trees. You have to do the privately owned trees too, because beetles have no respect for whether it's in somebody's garden or on council land, so you have to fund it and fund it 100 per cent," which they did.' To this day, the local council pays for the full cost of monitoring the disease, pruning or felling affected trees and taking other action to prevent spread, such as digging trenches between trees to prevent the passage of the fungus through grafted roots, plus removal and disposal of the wood. They are still experimenting with treatments, and in the centre of the town, Alister showed me some trees that he had inoculated with DutchTrig, a vaccine consisting of spores of a form of *Verticillium* fungus which causes the trees to wilt for a week or two but which seems to out-compete *Ophiostoma novo-ulmi* and thereby protect the tree from dieback. He'll spend his whole summer monitoring the many elm trees that have survived, following up on tip-offs from locals, and he can spot the onset of the disease from the first few wilting leaves. Then he will take a sample of wood to look for the characteristic brown staining under the bark that betrays the fungus. If present he will mark the tree, sometimes ringbarking the diseased branch to kill it to try to slow the fungal spread, then coming back to prune it after he has all the appropriate permissions. This can be the hardest part of his job and he regaled me with stories about the challenges of finding out who owns a property where he has spotted a diseased tree and then what to do if that person is in prison, or a Russian oligarch, or otherwise not amenable to their tree being tampered with. 'It's really difficult being the tree murderer,' he said. 'I've had people in tears.'

Where possible he will prune out diseased wood because if the fungus has not spread throughout the tree, it may be possible to eliminate it by just removing affected branches. The beetles tend to fly high and select feeding spots in the uppermost parts of the tree, which is why the first signs of wilt are usually spotted up there. He works down from that place. 'The rule of thumb is that you remove sections of bark to see the staining. Then you eventually get to that point, working down the branch, where the staining stops. And then you try to find another 10 feet below that point.' If he spots the disease early, on a big tree there can be plenty of scope for removing all the diseased material plus 3 metres beyond it, but on smaller trees or those where the fungus has spread more widely, sometimes the whole trunk has to come down.

After pruning or felling, Alister removes the wood to a store several miles outside of town. We headed out there and spent a while examining a mound of trunks and branches in varying states of decay – the detritus of one season's 'sanitation felling', the tree surgery he performs to extract diseased material. There has been a spate of disease in recent years – 700 trees were dealt with in 2023 – which Alister attributes to two things. 'Wood-burning stoves have become fashionable again so there's a lot of elm coming into Brighton. There's an awful lot of elm being cut in East Sussex as sanitation felling. So you've got infested wood that is coming in as firewood and then people leave it sitting around outside the back of their house for a year or two.' I took a mental note that back home we will need to be vigilant about firewood. The second cause is transport corridors into Brighton where diseased trees might not be so carefully handled as in the city. Yet even with several hundred trees a year succumbing to the disease, Alister is confident that it can be controlled enough to sustain the elm population. He also points out that diseased elms that were felled then regenerate, but once their shoots get to about 10 metres tall the trees can succumb to the fungus again. 'They don't require new invasions of beetles. They don't lose the infection. An awful lot of what we're losing has been felled in the past and regrown, but it was still dormant in the root system.'

I had set out the previous day from typical Scottish October weather, so I was overdressed for this very warm day – hot enough, it turned out, for elm beetles to still be active. After a bit of poking around among likely bits of bark with telltale holes, Alister said, 'There, look!' Trundling around on his finger was a shiny little brown beetle, barely the size of my pencil lead. After allowing itself to be photographed, it opened its wing-cases and flew off. I expressed surprise that they should still be so active at this time of year and Alister agreed. 'When I first came to Brighton in 1982 they were pheromone-trapping the beetles.' This involves catching beetles in small vessels baited with the chemicals given off by the beetles to attract mates or signal good feeding sites. Alister explained how this allowed them to track in detail what the beetles were up to. 'What they saw were two lots of emerging beetles in the summer months, because it's about six weeks from the beetle going into the bark and laying eggs to the next generation emerging. So we used to get two broods in the summer, and occasionally a half-brood later on in a hot summer, but now it's always three, and this is where it becomes a big climate change issue. Summers are that much longer.' He explained that the beetles only fly when the temperature is more than about 19 or 20°C, and they also prefer high barometric pressure. We speculated that this is a sign that the weather is improving, so they'll have a fine spell for their mating encounters. Clever little beetles!

Alister continued ripping bark off logs, showing me the distinctive brown staining caused by the fungus, the primary indicator foresters use to confirm a tree is infected, and also the beautiful doily pattern made by the brood-chamber and feeding passages of the grubs. This wood disposal spot is crucial because it concentrates the beetles and fungus in one area. Here the logs can be debarked and thus made uninhabitable to the beetles, or taken by a biomass energy company that chips and burns them, but even if they sit here for a while, the beetles that hatch find plenty of suitable habitat right where they are, instead of flying away to infect more trees. There are 'canary' elm trees about a mile from the wood store that he checks regularly to see if the beetles have spread out to them but so far the beetles seem content to stay where they are put, with its plethora of their favourite habitat of dying elms. 'They just go from log to log here,' said Alister. I fell to wondering where, at home, if the disease takes hold, we might stockpile diseased wood out of harm's way.

Alister took me back into town to catch my train, dropping me off in the centre where I could walk to the station via some of the most magnificent elms I have seen anywhere. I left Brighton inspired by their mastery of the Dutch elm disease problem. It has been extremely effective as a regime and the city is graced by many grand old elms shading the streets, giving structure and form to parks and sheltering the area's inhabitants.

Up in Scotland, there is another good news story. Edinburgh has also been successful in limiting the losses to Dutch elm disease. When the disease first arrived in the city it was prioritised by the local authority, and in the early 1980s they employed a squad of four people to spot early signs and take action. Rachel Blow was newly qualified in agriculture and forest sciences from Oxford and had moved up to Edinburgh for the festival season then landed a job as part of the team. She has never left. 'It was an incredible way to discover a city, climbing walls from one garden to another, walking through all the green spaces rather than by road. We used an axe to cut a nick in the bark to check for staining and a spray can to mark any diseased trees.' Edinburgh has no shortage of famous and beautiful green spaces and public gardens, including some very charismatic walks and streets lined with elms. 'In the early 1980s, the Meadows were 80 to 90 per cent elms, and they were everywhere in the New Town,' Rachel said. 'They were such a significant part of the aesthetic of the city.' Unlike in Brighton, the local authority didn't cover the costs of pruning or felling. 'We would mark the trees on large-scale maps and send the order to the landowner telling them it was their responsibility to deal with the diseased trees. They had to remove them and either burn them or strip the bark by 31 March of the following year.' She described an 'incredible row of ancient English elm' at Inveresk that the team had to order to be felled. Like in Brighton, experiments were carried out to try to save affected trees, such as injecting fungicides into particularly beautiful valued specimen trees in New Town gardens. But largely the disease control has been achieved through careful monitoring and felling of diseased trees.

Of course, prevention is better than cure, and reducing the spread of the disease is the best way of limiting casualties. The spores of *Ophiostoma novo-ulmi* are highly mobile and can easily be spread by people who think that they are being helpful. For example, if a chainsaw is used for cutting diseased limbs, fungal spores can remain on the tool or find their way into the oil used to lubricate the saw, and thus be transferred from tree to tree. These spores can remain viable for a week or more, so tree surgeons need to sterilise their equipment whenever dealing with infected wood.

Burning infected wood is a good idea to destroy the fungal sites, but if deadwood becomes firewood without being debarked, the threat it contains – either the fungal spores or the eggs or grubs of elm bark beetles – can be spread through the supply chain. In some rural areas firewood, particularly hardwood like elm, can be transported considerable distances. In my part of Scotland there's strong evidence that firewood has been a powerful vector of the disease. The Highland Elm Project at the Scottish School of Forestry has been taking a detailed look at the northwards movement of the disease, tracking surviving elms and analysing their genetics. This research shows the disease spreading along the road networks and over distances bigger than the mile or so that the beetles can fly unaided.[11]

Another interesting British case study is the Isle of Man, in the Irish Sea, which has an estimated 300,000 elms.[12] Dutch elm disease was first noticed there in 1992, but by 2016, only about 2,500 trees had died, less than 1 per cent of the total population, compared with up to 99 per cent in other parts of the British Isles. The surviving trees included an elm clone that is known to be highly susceptible, so natural resistance has been ruled out as an explanation of the low incidence of blight. Crucially, the climate of the Isle of Man is marginal for the elm bark beetle, with generally cool summers and lots of wind, thus limiting the resident beetles' opportunities for gathering and spreading the fungus. An 'island effect' means that it has been possible to reduce risk of new incursions via firewood or infected beetles flying in from nearby areas, plus there is a rigorous and effective regime of sanitation felling whenever the disease has been noticed. It's clear that even when Dutch elm disease can't be completely avoided, a lot can be done to prevent the tragic losses of the past.

The Crocach Sisters

In this part of the world, the nineteenth century is notorious for what became known as the Highland Clearances, the enforced removal of people from their homes to make way for sheep. In the north of Scotland, landownership has been in the hands of a shockingly small number of people for centuries, and 200 years ago, the landowners took control, for their own exclusive use, of land that had been managed collectively and treated as a local commons for generations. In the Highlands the landlords claimed this was a form of 'improvements' of conditions for local people, so they were forced to find more purposeful jobs than subsisting on the land. Here in Assynt people were forcibly cleared from the inland glens, where sheep farms were set up in their place, and pushed into small parcels of poor ground along the coastal strip, where they were supposed to 'benefit' from access to work in fishing or seaweed collection. The result was dire poverty, mass emigration and people forced to survive on welfare projects. One of these projects was the back-breaking construction of 'peat roads' – built tracks into areas of moorland where people cut peat turfs in spring, laid them out to dry during summer, then carted them home in autumn to burn in the winter.

The nearest peat road to my home is called the Crocach Road, because it leads from the crofts at Richarn out to a large and stunningly beautiful, intricate lake system called Loch Crocach, which is surrounded by a huge area of boggy, rocky, lumpy, bumpy land that is our township's common grazing. After the local crofters bought the land from a private landlord into community ownership in 1993, our grazing committee established a native woodland plantation, so the Crocach Road is now mostly in the shelter of birch, rowan, alder, willow, pine and, under Craig Darroch, a few oak trees. No elms have been planted, and I'd never considered the area to be elm territory, until recently.

At the end of the peat road, there is a vast expanse of land and water, which runs for miles out to the foot of the mountain of Quinag. This area is completely uninhabited, but there is a ruin of a house just above the shore of the loch, with a magnificent backdrop of the characterful Assynt mountains lined up along the horizon to the south, and to the west, the blue-grey ruffled blanket of the Minch, our sea channel between the mainland and the Outer Hebrides. Around the ruin, there are signs of old fields, a trace of what was perhaps a holding pen for livestock, a tantalising suggestion of another small building beside a stream running down to the water's edge, and a few isolated trees. Some scrubby birch and alders cling to shore-side rocks. I pass two cows: mother and calf, wary but not frightened, watching with their curious, shining, mournful eyes, backing away as I approach with that beguiling cattle standoffishness.

Nearby are two slender elms, standing well apart from any other trees. They are a pair, and their delicate forms suggest something feminine. Sisters, I think. They may in fact be growing from a common rootstock. From under them, there is a perfect view across the loch to the distant mountains. It is a cold winter's day, but bright, blue and beautiful. There is no wind and the trees are quiet. The loch water is deep navy blue, the colour of bleak contemplation. We have an abundance of land, but so much of it is empty: houses where families used to live are crumbling into the heather and no one these days remembers who belonged there. Did someone plant these elms back when the cottage had peat smoke wafting from its chimney? Or are they a remnant of some earlier natural woodland?

I return to the sister elms in late spring, when the seeds are ripe on other trees, but there are no seeds on these. I can see now that they are storm-battered: their upper branches are dead, but they don't seem to be showing signs of disease. They are just old and exposed to fierce elements. On their bark is a blossoming of lichens of many species, signs of the clean air and the high annual precipitation levels (more than 1,500 mm) that qualify our woods as temperate rainforest. They are presumably too far away from any other elms for their flowers to be pollinated. Neither my sister nor I have children, though through marriage we have both become part of large families: the seedlessness of these two seems appropriate. Their battered appearance too – we've both been through stormy times recently. Around them is a mix of woodland and heathland flowers: bluebells and heath speedwell, both as blue as the sky was when I visited in winter; pignuts and heath orchids, white like today's bright clouds; yellow pimpernel and tormentil like the sun that's trying to come out. The base of the trunk is about to be engulfed in a summer flood of fronds of lady fern, hard fern and the ubiquitous bracken. If there were more trees around, the bracken would be shaded out, but in open ground like this it dominates. In a month or so, it'll be head-height and getting out to these trees will become a battle.

It is common in this part of the world to find rowans standing alone in the landscape. Some of these are clearly the result of a random bird dropping a seed where the teeth of herbivores can't reach – sometimes they are in really unlikely spots on crags or big rocks. But tradition states that a rowan tree growing close to your house will protect you from evil spirits and there is a taboo against cutting them down as it is supposed to bring bad luck. So some areas that must have previously been woodland or scrub and have been cleared by crofters for grazing livestock have no trees left except for the rowans. These elms are standing alone in a similar way: two tender-looking trees, exposed to all the forces of the weather with no willows, birches or hazels to give shelter. Were they deliberately not cut when everything else was felled for construction or tools or firewood or just clearance for sheep? Does their avoidance of the axe suggest a similar taboo or belief in their power of spiritual protection? Or did they have a greater value left standing, perhaps for fodder, or for their medicinal use or as starvation food? Were they planted and protected by someone with aspirations of parkland grandeur or who simply thought they would look good in the view? There's no one left who knows.

These elm sisters stand as a testament to the lives of the people who endured so much hardship here in the past but who have been lost to history. And stand they do, in the teeth of gales and storms, enduring beyond human memory. The poet

Norman MacCaig wondered at the end of his long poem 'A Man in Assynt', whether the 'ebb, / that sad withdrawal of people, may . . . / reverse itself and flood / the bays and the sheltered glens / with new generations replenishing the land'. These days we own this ground and there are woods we planted growing now within a stone's throw, offering hope, perhaps, of a more sheltered future, healing some wrongs of the past and restoring the land.

Chapter Five

Death: Coffins and Cartwheels

In August 2023, everything began to unravel in 'this most beautiful corner of the land', as Norman MacCaig put it. While I was visiting a neighbour to discuss a knitting pattern, she told me the grand old wych elms growing by Glencanisp Lodge, one of our community buildings, were succumbing to Dutch elm disease. One had already been felled by the ranger as it was in danger of falling on visitors using the path into the mountains. My local elms were not, after all, safe.

The next morning I made my way to Glencanisp and tracked down the ranger, who confirmed the sad story. Standing by the wreckage of the felled tree, I realised that our last refuge from the disease was no more. My blight-free haven was infected. It felt like the day COVID-19 arrived here after many months when we were one of the only places left in the Northern Hemisphere to escape it. The sky didn't fall in, but there was a deep sense of something subtle having dropped away. Hope.

Across the road from the big elm at Torbreck I wrote about in chapter two is the old manse. I've known this house for more than thirty years, since long before I lived in the area. The Vestey family, who made a fortune from butchery and meat canning and spent some of it on a Highland estate covering the bulk of our parish, sold the inhabited part of the estate in the late 1980s; when their buyer went bust, the land – marketed as a place where people were 'aliens in the landscape' – ended up in the hands of a Swedish bank. The 130 crofting families, most of whom had been in the area for many generations, did the seemingly impossible to raise the very large sum to purchase the land and set up a crofter-owned and controlled company that is our landlord to this day. Apart from the crofters' houses, the only building on the land with any commercial potential was the old manse – the Church minister's official residence, at Torbreck – and this was sold to a small software company to help raise the purchase money.

Meanwhile, I was at the University of Edinburgh. My PhD in Artificial Intelligence took me to fascinating parts of the world, working with land managers, especially foresters, who were keen to try out experimental software. One of them was Bill Ritchie, the architect of the crofters' buyout (and these days, my husband). I had already got Bill involved in a software trial and was visiting the area regularly, and it wasn't long before the new company based in the old manse became embroiled with us in a project, along with several partners around Europe, exploring how computing advances might help us protect the environment. In the early 1990s I attended frequent meetings in the old manse, walking or cycling up the drive flanked by its line of eight magnificent elms. Now I feared they were at risk of going the same way as the Beauly Elm.

So, it was with a mind full of memories that I approached the manse to ask to take a close look at the elms. I knocked on the door and had a pleasant chat with Grahame, the current owner, who gave me permission to inspect them. Most of the eight are huge, with ash, sycamore, hazel and spruce growing among them – the hazel and ash probably naturally, the others more likely planted. I was pleased to note no dieback on the ash as infection with the Chalara fungus is ravaging so many around the country. I made my way from trunk to trunk of the lovely big elms, soaking up their shade and stately atmosphere.

One of the trees directly below the house, a vast, multi-stemmed being, had a trunk blow down in a storm last winter, but it was very much alive, lying horizontally or even below horizontal where it had fallen down the slope, all the side branches now upper branches, in full leaf and growing strongly. Its other trunks too appeared healthy enough, though rot in the trunk may have caused the fall, and maybe this was a sign that all was not well.

A cursory glance at the second in the row suggested my concern was in order. The upper branches of the tree were full of wilted leaves. The rich green of a healthy tree had turned greyish, the lush foliage limp and shrivelling. This tree looked sick.

Although I've spent thirty years working on trees, I'm not a forester or a botanist, not a silviculturist or a tree surgeon. I'm a tree-hugger, for sure, but my scientific qualifications are computer-based and my experience is in modelling, campaigning and being creative about trees, not in diagnosis of their diseases. Still, as a passionate layperson with an elm obsession, it was pretty obvious that the tree was dying. My heart sank into my boots.

Was Assynt, the last refuge against Dutch elm disease, also now succumbing?

Realising all I could do was treasure the trees as long as they're standing, I listened to the rustle of the leaves, the twitter of some great tits, a wren scolding, and enjoyed the dappled light below the graceful boughs on this warm August afternoon. If these trees should die, what a gap it would make in the woodland texture of this glen. The building that has been granted privacy and shelter, gently framed by the trees, would stand exposed, a stark transformation like the one suffered everywhere elms have been lost. As this is the manse, it felt appropriate to pray, to some tree deity perhaps, to protect them.

Only a few days later, my sister called to say our father had contracted a chest infection. He was old and feeble, living in a care home, years into the slow advance of dementia, unable any longer to recover from a series of strokes, the most recent of which had robbed him of much of his power of speech. Should I drop everything and go to Glasgow to see him? No, she thought, he'd probably last a little while longer. He was dead before I got there the following morning. It initially felt a 'release'. Nobody wants to die slowly, faculties and organs decaying little by little, incapacity creeping inexorably on, enjoyment and sentience becoming increasingly fragile and tenuous. Now he was gone it was possible to see his life as a whole, to try to place him in memory as the person he had been before dementia took him. After my statutory four days of bereavement leave I carried on working, but with a sense that something glacial in me was calving, that something monumental and foundational had fractured. When it broke, the impact was shattering.

Memories came thick and fast. Here's one. There were badgers in the elm wood close to our Northumberland home and our father used to pride himself on knowing where they were and how to spot them. He would take my sister and brother into the woods at dusk to sit by the sett and wait for them to emerge. There was a family myth, which even made it into one of the eulogies at his funeral, that he didn't take me because I wouldn't sit still. In fact, what had actually happened was that he led me to the sett one evening and we sat down close to an elm tree. An owl was hooting in the canopy. After sitting in the dusky hush, I noticed, perfectly camouflaged at the base of the tree, an owlet. It was fluffy and blinked slowly as if wishing, by closing its eyes, that we would vanish. Presumably its mother in the treetop hoped the same thing. I knew exactly how they felt. I pointed the fledgling bird out and we backed away, quietly. No badgers that day, though I saw plenty on my solo rambles. Sometimes I would have to rip wire mesh out of the sett entrances, stuffed in there by badger-baiters to prevent escape when they sent dogs in. My relationship with my father was traumatic and in one way I didn't mind that he branded me too fidgety to be taken to the woods – I much preferred going on my own anyway.

As time went on, throughout the woods, the ghostly forms of once-living trees created their own magic. My elm stump remained in some way present as a tree, its former grandeur perpetuated in my memory. I used to sense elf presences among the dead elms. More than mere skeletons, or corpses, they proceeded to shelter owls and woodpeckers and to soften with mosses. Life grew in and around them, continuing on. And yet they haunted the woods, just as our dead don't really leave us and insist upon finding things to remind us of them, and as the bark rots away and as our mourning goes on, they reveal what they were really like inside, ugly though that may sometimes seem. The next few months are best passed over, other than to say that even when I couldn't work, couldn't speak to my friends, wasn't sure who I was anymore, I consistently found solace with elm trees. I set about visiting all the elms in the parish and each visitation held meaning or significance: a small tree squashed by a huge fallen trunk spoke eloquently of how I felt in the face of my father's death; a tree hanging halfway up a crag showed me life's precarity; apparently healthy trees threatened by this invisible disease mirrored my own physical wellness that masked the uncertainty of my survival. At times it felt as if the elms were helping me even in the utmost extremity of distress.

Haunting is rarely welcome, though it's a way the world holds on to truths, so it's unavoidable and fascinating. And while ghosts may be slow or difficult to lay to rest, life goes on. When trees or people die, their absence creates space for the past to be present. Death enables a kind of time travel by returning former times to us, and sometimes, I am discovering, after a death, old wounds can be re-opened and cleaned and allowed to heal.

Elms are valued by people for their leaves, bark, flowers and seeds, and their significant landscape value has only been properly appreciated as it has been lost to disease. But there can be little doubt that the greatest economic value of an elm tree comes when it is cut down, because its timber is so prized. Woodworkers appreciate its distinctive grain and extraordinary patterning when burred, and it is also favourable for carving and turning. When green, it is waterproof. It is rot-resistant when wet and above all it is strong.

Furniture makers are drawn to elm because the grain of the wood is often beautiful, and woodworkers have given it many lovely terms: 'partridge breast', 'grain and flower' and, in French, *'pied de chat'* ('cat's paw'). Robert Somerville, who has worked closely with elm wood in the construction of whole barns, says, 'It is as if the grain is rolling, churning over itself. It eddies and swirls this way and that, as if directed by unseen forces beneath the surface, like water over boulders in a river.' This swirling pattern is caused by the xylem and phloem fibres that form the sapwood not growing straight up but spiralling around the trunk or branches, and doing so with varying direction and thickness in different years. This cross-grained effect is not merely decorative: it makes strong pieces that can withstand serious force.

When elms started to be felled in the south of Scotland in the 1970s and 80s, as a result of the arrival of Dutch elm disease, the presence of large amounts of elm wood led to a renaissance in furniture building, led famously by the Woodschool workshop set up in Jedburgh, in the Scottish Borders, by Tim Stead, whose eye for the natural patterns in the grain gave his pieces the look of artworks while remaining fully functional. One of my favourite kitchens in the world has a Tim Stead elm table as its centrepiece. Sitting at it, tucked into a nook in its welcomingly irregular circumference, elbows leaning on its warm surface, looking into its swirling grain, is as comforting as stirring a pot of wholesome, delicious soup.

Elm produces burrs, which are formations caused either by the tree responding to insect activity or stress, or in the case of wych elm, just because it is something even a healthy tree does. The resulting cluster of knottiness is, to some, a kind of deformity, but to a wood-turner it is a gift of rich patterning. Elm burrs have gone in and out of cabinet-making fashion over the centuries, but the resulting pieces are often stunning. Veneers made with burrs can be used to achieve extraordinary symmetrical effects. With a chunk of burr, a wood-turner can create beautiful bowls and other vessels that show off the complexity and range of colours created by the tree. I have an elm burr fruit bowl, a precious gift from my mother, which will last my lifetime and probably many more, and without any staining or polishing it retains a spectrum of shades from nearly gold to almost black, a galactic complexity of little spirals and nicks with a central three-way swirl of grain that begs to be stroked, and a gleaming outer surface that looks like glowing embers and licking flames. I think this is my favourite kind of art – work that emerges from wonder at the natural world, where the artist shows what has grown of its own accord, engaging with rather than imposing on nature, showcasing a harmonious partnership between humans and trees.

Here in Scotland, one of the earliest wheels ever found by archaeologists, more than 2,000 years old, has an elm wood nave, or centre. But more than a thousand years older than this, the earliest written reference to elm dates all the way back to the Mycenaean era, from the inventory of stores in Knossos, in Crete, where chariots had elm wheels. In the nineteenth century wheels were still being made with elm naves and sometimes also the felloes, or rims, of wheels were made from elm. H. L. Edlin describes why elm is the perfect wood for this purpose, along with other applications that need to take regular punishment, like wheelbarrows and mallets: 'You can drive a wheel spoke or a chair leg into a piece of elm, and it will be held firm, whereas wood of nearly any other kind would be rent asunder.'[1]

The author of the 'elm bible', R. H. Richens, explains elm's extraordinary strength is due to the way the fibres of the wood interweave and lock together so that it is cross-grained, giving it elasticity but also making it resistant to splitting.[2] This means that the wood can withstand severe impacts, which is why it has been used in mine and quarry trucks, and for the heads of mallets. Elm is perfect for some jobs, such as timber framing of buildings, which requires huge mallets, called 'beetles' or 'persuaders', particularly what folklorist G. E. Evans describes as 'a strong, knurly, knotted and crooked' sort of elm found in Essex, the tools from which 'with using wear like iron'.[3] Carts and wagons that needed to survive people shovelling materials onto and off them were often made from elm boards. Traditionally the best wheelbarrows had elm wheel hubs and bodies made of elm boards, and an English specialist manufacturer, Stubbs, still makes them according to a 130-year-old design, though these days with a metal wheel with pneumatic tyre. If this book sells well, I'll be acquiring one, confident that I will be shovelling horse manure into it for the remainder of my life!

In timber-building designer Robert Somerville's brilliant book, *Barn Club,* he narrates the entire process of elm-barn construction, from choosing the trees in the wood to the final topping-out ceremony, and in the course of this year-long project it becomes clear that a crucial factor is to use the wood while it is still green, when it 'cuts like cheese', and not to wait for the cross-grained fibres to lose the moisture of the inside of the tree and lock hard together.[4] Waiting for wood to dry out and become 'seasoned' can ensure that the resulting wood will not shrink or change shape; however, in the case of elm it makes the wood much more difficult to work with.

There are other traditional applications that require the toughness that elm can deliver, some of which are peculiar. Ploughs are one of these, with elm having been a preferred material for the 'mouldboard', which is the big slab next to the sharp 'share' that cuts into the ground. The mouldboard does the heavy work of turning the soil, so elm's strength is ideal. From the iron age, plough shares were forged and in many cases the entire apparatus was made from iron. However, metal has disadvantages because a claggy soil like clay sticks to it, building up and making hard work of the job, whereas a smooth piece of elm wood sheds the lumps of dirt and is therefore preferred on heavy soils. Other agricultural uses include situations where heavy livestock like cattle or horses are likely to be kicking or bashing around, such as dairy or stable partitions, cattle feeders, hay cribs and so on. Similarly, for heavy-duty bread making, a kneading trough for dough was ideally made from elm, as were bellows for fires, especially ones that required a lot of use, such as those belonging to blacksmiths.

The large size of mature elms means that the wood is often used for projects needing big slabs, such as what Colin Tudge describes as 'the buttock-moulded seats of rural armchairs', the makers of which were known in some places, appropriately enough as 'bottomers'.[5] Fiona Stafford waxes lyrical about the results: 'A polished elm seat is often so carefully crafted that the lines seem to spread from the central rise like peaty water rippling away from a smooth stone. There is no need for a cushion if the bottomer got things right.'[6] The wood also has antiseptic qualities that have made it particularly effective as toilet seats.

Even older than elm wheels are elm weapons. All manner of spears, swords and shields have been dug up by archaeologists from the Neolithic period (New Stone Age) around Europe, and even older bows of elm have been found from the Mesolithic period (Middle Stone Age) in Denmark and Germany.[7] Bows were a crucial technology for hunting as well as warfare, traditionally made from yew or elm because they have good elasticity as well as being strong enough to resist the incredible force of the tightly drawn string when it fires the arrow. Some argue two Celtic tribes in ancient France took their names from these two trees: the yew warriors formed the Eburovices and the elm warriors the Lemovices, from which Limoges gets its name.

As well as being practical, bows were part of the paraphernalia used in ceremonial burials. Perhaps it is no surprise, therefore, that elm, alongside yew, has deep cultural connotations with death and the afterlife. Elm is the traditional first choice of material for the boards of wooden coffins due to its durability in wet ground, and these days, with big elm boards being harder to source, it is often used as a veneer on top of cheaper wood like pine. Another link with death is elm's use for gunstocks, again because of the ability to withstand the recoil. Used from the earliest mediaeval guns, this created a military imperative for the reliable supply of elm timber. According to Richens, 'The planting of elms along trunk roads in France created a highly distinctive landscape feature. In origin, it was a measure to supply the royal army with a necessary item of ordnance.'

All of this has made elm wood particularly valuable. Penalties for cutting a tree down were meted out to common people by the rich and powerful, who owned all the land. As the traditional Scottish rhyme puts it:

> The aik, the esh, the elm tree,
> The Laird can hang for a' three.

In other words, someone felling an elm without permission, along with other prized timbers such as oak and ash, could face the ultimate penalty: a death sentence.

Elm wood's waterproof properties mean it has played a key role in the development of boats and is still sought after by boat builders. It was particularly prized for the garboard strake, the lowest plank of wood or wale – as the joke goes, not to be confused with the whales, who are down below – next to the keel of wooden ships, where strength and rot resistance are key because it is both deep in the water and also likely to have water swilling around inside the bilges. One of the last and fastest tea clipper tall ships, the *Cutty Sark*, was built using rock elm for most of the wood below the waterline. Elm was used for many other parts of sailing vessels too: keels, stems, ribs and floors, as well as for applications where it was the toughness rather than waterproof quality that mattered, such as pulley blocks, dead-eyes and trawler rollers. A big elm burr is an ideal mast-step, the chunk of wood that a mast stands on, as it must absorb huge forces. Canadian rock elm (*Ulmus thomasii*) was particularly prized for structural boat timbers as it is as strong as its name implies and it can be curved into shape without splitting when steamed. I was excited to learn that such steam-bent rock elm timbers were used for the model of boat my husband and I sail, a Twister 28. That elm trees have contributed to all the joy I've gained from sailing over the years literally floats my boat!

There are many other boat-related uses: paddle-steamer paddles were sometimes elm, it has been used for bailing buckets and it's a good choice for oars. Anyone familiar with Patrick O'Brian's classic nautical novels set in the Napoleonic era will no doubt delight in the captain's steward, Killick. He is named after a traditional Cornish anchor, which consists of a big stone held within an elm wood framework to which the anchor rope or chain can be more securely attached than to the stone: the perfect name for a character who is both cunning and deeply dense, but who keeps the captain grounded.

Waterproof elm is found in water mills, sluices, pumps and as pilings in harbours: much of Venice was built on elm pilings. A 1970s survey of the utility of elm notes that 'a study of the deterioration of structures in seawater in 1920 reported that rock elm, which was used for the sheet piling of dams in which the foundation of tidal works was laid in part of the Tyne Dock, was sound below beach level over forty years after it was laid.'[8] The same survey reports that when London's Waterloo Bridge was demolished in 1936 it was found to have been bedded underwater on elm timbers, which were still sound after 125 years.

The most intriguing exploitation of elm's waterproof property is when elm logs are hollowed out and used as pipes that can be trusted not to leak or rot even when buried long term, which explains their widespread use for water mains. For those of us living in the era of universal plastic, it is amazing to think of serious metropolitan plumbing using wood for water pipes, but it makes sense: most wood, elm included, is non-toxic, unlike metals such as lead; wood has more flexibility than ceramic, so is less likely to be crushed or snap under loads or if there is a ground shift; and it can be whittled and cut to make joints. There are records of elm water pipes in London from the thirteenth century, and in the seventeenth and eighteenth centuries it was the go-to material for drains and for transport and piping of water, especially in England, where elm was widespread in hedgerows and woods. According to the Woodland Trust, London, Bristol, Reading, Exeter, Southampton, Hull and Liverpool all had elm water mains.[9] It was also used in the USA, where new towns and cities had water mains installed from the early nineteenth century; Philadelphia had 45 miles of wooden water piping, some of which was hemlock or pine, but much of which was elm. But this was all old-hat to the large Dutch population of Philadelphia – the Dutch, a source of much water-engineering innovation, had already been using elm for piping water for 2,000 years by then: near Rotterdam, archaeological finds of wood pipes used in sluices for diverting water away from dams, built to block high tides from inundating coastal areas, have been dated back to 70 BC.[10]

To achieve the ideal wood for a pipe, elm trees were managed in a particular way. According to seventeenth-century reports, all of the side branches were trimmed off the growing tree, leaving only a small canopy. Landscape paintings of the time, such as Jan Siberechts' *Landscape with Rainbow, Henley-on-Thames*, which he painted in around 1690 and is displayed at Tate Britain, show many of the resulting tall, slim trees with tufty tops. This practice of trimming off side-growth persisted long after there were alternative materials for water pipes because the tall, straight poles that it produces are useful for so many other purposes.

To make a water pipe, the trunk must be hollowed out. It is then cut into lengths of about 2 metres, then bored out using a large auger, a metal tool with helical blades that are shaped with a twist to pull out the material as they are driven into the wood, in the manner of a corkscrew. Each section is tapered from one end to the other so that the narrow end fits into the broad end of another pipe. Originally this was all done by hand, and the grunt, rasp and squeak of wood boring was a feature of elm wood mills. It's no doubt tiresome and repetitive – the origin of the more usual sense of the word 'boring'.

It is somehow intuitive that given elm's watery links, it doesn't make good firewood: it's hard to light and it doesn't burn well. 'The Firewood Poem' by Celia Congreve includes the lines:

> Elm wood burns like churchyard mould.
> Even the very flames are cold.

In Robert Graves' *The White Goddess*, fire from elm is even more reluctant:

> Elm logs like to smouldering flax –
> No flame is seen.

Nonetheless, plenty of people are duped by the 'good hardwood logs' sold by firewood sellers who get hold of the large amounts of deadwood generated when Dutch elm disease strikes. Keeping elm wood in your wood store is a really bad idea, both because it fails as good firewood and its bark is the ideal habitat for the devious little elm bark beetles. However, it is an excellent choice for a Yule log, which is burned variously at the solstice, Christmas or Hogmanay fire. A Yule log should burn for a long time (to maximise time for social activity) and leave enough to light the fire the following year, as a good luck charm to keep continuity from one year to the next. Elm's reluctance to burn makes it ideal for smouldering away all night long, and the resulting charred log will be guaranteed to have no fungus-harbouring grubs left in it.

This slow burn can be helpful in other contexts, and the indigenous Arikara people in North America used it for gently firing pottery. It makes poor charcoal, though mixed with other wood it is sometimes used for production of barbecue coals for slow cooking.

Elm's hardness and reluctance to spark into flame also makes it useful for applying brakes on windmills, where it faces tremendous forces that generate heat. Blacksmiths' blocks were also often made of elm for the same reason. Another example is in the pre-matchstick striking of fire, using a drilling stick and block of wood to generate friction to set tinder smouldering. In Japan, the Ainu people's word for elm is *'chikisani'*, derived from their word for fire-drilling, *'chickisa'*. Other Indigenous people who used elm for fire-drilling include the Iroquois in North America.

After elm has finally been made to burn, the ash that results is high in potassium, so it makes a useful potash fertiliser or the basis of soaps. This perhaps explains the tradition that once the Yule log has finally burned away, its ash is scattered on the earth to ensure good harvests the following year.

Elm is less dense than some other woods, such as oak, and therefore preferable for uses where strength is required but weight is a disadvantage, such as tool handles. In the Highlands, elm makes a perfect caman, the stick used for the game of shinty. The caman is more than a piece of sports kit, it is a cultural icon. While an oak caman is robust enough to be effectively unbreakable, the 'sting' of impact from the hard, wooden – often hazel wood – ball, can result in injury. Birch wood is flexible and light, but can easily break. Elm is both strong enough and light enough – just right, like Goldilocks' porridge and bed. Ash wood, it must be conceded, is perfect too.

There are sports that need an even tougher stick than shinty – ice hockey, for example, and in Canada, where it was born and is the national winter sport, an elm grows that is even tougher than average. Rock elm is aptly named for the hardness of its wood, and it makes incredibly robust hockey sticks that won't crack or split even in seriously cold temperatures. For those who prefer their sport indoors, elm makes an excellent dartboard, accepting the points of the darts but not splitting when they make impact.

There are situations where elm is unsuitable. Its wood can have a tendency to twist, which means that it's avoided for precision jobs, such as musical instruments, unless they need to be really tough, such as percussion instruments and in particular bells, which can cause huge strain when rung – especially big church bells, which need something powerful to swing from.[11] Elm was the traditional material for the 'headstock', the heavy block of wood from which the bell was suspended. Many churches furnished with bells from the mediaeval period onward still have their original elm fittings. So at your funeral, not only may you be buried in an elm coffin, but when the bell tolls you farewell, it may well be hanging from a big chunk of elm wood.

There is a tiny poem, just a couplet, usually claimed to be written by that over-credited and mysterious poet, Anonymous, although it also turns up in a poem by Kipling, who may have composed it or taken it from the oral tradition. It goes as follows:

> Ellum hateth
> Man, and waiteth.

It's a strange little ditty. What does it refer to? There is much debate. Some say it refers to elm being used for making coffins, so an elm tree is watching us in preparation for our mutual burial. Another theory is that, as the great forest writer H. L. Edlin puts it, 'the wicked elm waits for its victim to sit beneath it', referring to elm's strange habit of dropping heavy chunks of wood on unsuspecting people lingering under the tree. It is not properly understood why elm sheds branches, but it does. We're not talking about limbs being ripped off by winter storms, but an apparently healthy elm tree simply dropping a whole branch on a calm summer day, without warning. It is thought that perhaps what underlies this strange behaviour is that a winter storm has caused some structural damage internally, not visible from below, and when the branch has to carry the full weight of dense summer foliage, it suddenly gives way. This seems more plausible than the idea that it has murderous intent, but who am I to say?

In the forest environment, falling branches are far from a problem, and indeed a tree, over the course of its life, will shed a vast amount of leafy and woody debris from the canopy. Such 'litter' is a vital part of the woodland ecosystem, providing food and homes for myriad other life forms. The wood of the tree itself also hosts other creatures, with a portion quite possibly rotting away while the tree still lives – big old elms can be dramatically hollow, as we saw in chapter two, and all the heartwood will have been consumed by insects or fungi, or chipped out by birds or bats or even bears. Certainly, once a tree dies, its body provides an ongoing feast for other species, first while the tree is still standing and then when eventually it falls.

We humans tend to have an aversion to the ebbing part of the lifecycle, when the tide of life flows out, and our language around it is generally negative. 'Deadwood' refers to people who are no longer serving any useful purpose. The name for a standing dead tree is a 'snag', also the word for a problem that should be got rid of. In forests, woodlands, parkland groves and everywhere trees are grown and managed by people, dead trees are viewed as unwanted or even dangerous, and so they are felled. Deadwood that has fallen is considered messy and tidied up, taken to a dump or chipped or burned. Trees that fall in rivers and other watercourses are dragged out as impediments to the free flow of water. It is an exceptional forest manager who welcomes the presence of deadwood in their environment, or considers it to be an important part of the ecosystem, but it turns out that this is just the kind of management we need.

We need to treasure deadwood. It is home to an astonishing amount of life, and removal of it from a woodland threatens or eliminates that life. It also has huge climate change implications as deadwood stores the carbon that a tree has absorbed over its lifetime and gradually releases some of it, as it rots down, back into the atmosphere while passing the rest of it down into longer-term storage in the soil, locking away some of the carbon dioxide that would otherwise act as a greenhouse gas. Far from being dead, deadwood is a hive of life, and far from being problems, 'snags' are a key part of the solution to our biodiversity and climate crisis.

Deadwood is a mass of fibres and material of varying densities and textures, and when it rots, this variety increases. As it softens, due to the action of fungi, which are the first and primary force in the decay process, the wood becomes a food source for all sorts of creatures, who either eat it directly or eat the fungi, and it even becomes a source of shelter. Lots of beetles are able to chew into the hard outer layers of a log to create nest sites and to lay eggs, and when they hatch the grubs feed on the softening inner wood. They are known as saproxylic beetles – from the Greek *'sapros'* for decay and *'xylon'* for wood, the same root that gives us the wooden musical instrument, the xylophone, and also xylem, the vessels in the layer just under the elm trees' bark that block up in response to the Dutch elm disease fungus. Being a rotten wood specialist is an excellent strategy for a beetle, one of its great advantages being that it can hide and lay its eggs out of sight from predators, so it's widely adopted. Around two-thirds of all beetle families include at least one saproxylic species, and there are more than 600 saproxylic species in the UK alone. France has more than 3,000 species. Exactly how many there are in much-studied Europe is, according to the IUCN Red List, 'unknown', so you can pick a five-finger number out of a hat for the global total, especially as diversity will be highest in the tropics, where there remain vast numbers of as yet undiscovered species of insects.

Deadwood also provides shelter for birds and bats. It is estimated that 30 per cent of Europe's birds depend on deadwood for nest sites or food sources – by no means all those beetle grubs remain unmolested – and many birds, including woodpeckers and owls, make their homes within standing dead trees.

We don't tend to think about fish and trees together, but fish benefit from deadwood too, and from trees on riverbanks, streams and burns, lochsides and other watery places where elms flourish. In my own parish, the steep-sided gorges and ravines are where many of the elms cling on, safe from the teeth of herbivores and the ravages of hill fires. Their shade cools the water in hot weather and reduces the chilling impact of winter winds, so they buffer out extreme temperature changes. Their leaves and twiggy detritus, as well as lichens, epiphytic fern fronds and so on, drop into the water below and feed the invertebrates that fish live on. When bigger pieces of wood fall – those unexpected branch drops – or indeed when a whole tree topples over, it can slow the flow of water, creating safe pools for fish and reducing erosion of riverbanks.

Land managers, especially in public places, tend to want to keep things looking neat. Deadwood looks like mess, it clutters up our view, it's disorderly and chaotic, but we need to resist the urge to clear it away. In his book *The Insect Crisis*, Oliver Milman advocates for the many insect species who love the random messiness of rotting vegetation and are thwarted by our incessant tidying up.[12] He urges us to help insects by indulging in indolence and simply doing less: less weeding, less mowing and especially less clearing away of the so-called 'litter' dropped by trees. Bugs will benefit and so too will all the other animals that eat them.

When Dutch elm disease strikes, the large amount of deadwood creates a mini-biodiversity boom. Knowing that urban environments have had success in limiting the disease by the removal of dead elms, I worried about leaving dead elm wood in the forest, but Max Coleman reassured me: 'The wood is only a viable habitat for the bark beetles for a couple of years, so the risk is short term, whereas the benefits of the deadwood in the ecosystem lasts for decades.' If we are willing to take a long-term perspective and widen our view, leaving dead elms in the woods would be a boon for biodiversity – one sort of loss can be another kind of gain.

The Edinburgh Wych Elm

The Royal Botanic Garden Edinburgh is one of my favourite places on Earth. I was first taken there by my mum, then I forged my own relationship with it. My parents, though scientists, were self-confessed culture vultures, and every summer they went to Edinburgh for the amazing festivals that take over the city during August. Mum booked a student flat in one of the university halls because it was relatively cheap and walking distance from the Usher Hall, with its big orchestral concerts; the King's Theatre for plays; the Meadows, with paths lined with trees, many of which used to be elms; the Royal Mile, with all its free entertainment, buskers and Fringe performers; and the chamber concerts in the Queen's Hall. I joined them as often as I could – it was my favourite family tradition. My father went in search of opera singers and comedians, Mum and I to see dancers and, often, to 'the Botanics', where we'd have lunch in the hilltop café then stroll around enjoying the exceptionally rich collection of trees and other plants, calling in on her favourite tree ferns in the glasshouse. She had deep knowledge of botany, and I always learned something from her, but mostly I just loved the lightness of her mood when surrounded by the beauty and variety of plants. She often referred to her dad's dad, who had been a gardener for a big house in Yorkshire, and her happy memories of playing in the greenhouses and formal garden he cared for. In the summer of 2013, I was poet in residence in the Edinburgh Botanics, and I don'think she was ever more proud of me!

When Dutch elm disease struck Edinburgh in the 1980s and the elm trees started dying, the Botanics played a key role in the city's response, one of the most successful in the UK, with tree surgery, including root severing, and prompt felling of diseased trees limiting the spread. But the beetles will fly and, inevitably, a magnificent elm in the Botanics itself succumbed to the disease. Rather than simply cart it away, staff devised a plan both to celebrate that tree, and wych elms in general, and to raise awareness of the impact of the disease. The tree, described by Max Coleman as 'majestic' and 'a dominant presence', and by Ian Edwards as a 'giant' and 'prominent landmark', may have been the oldest tree in the garden – its rings showed it to be 197 years old when it was felled in 2003, and it therefore pre-dated the establishment of the garden on its current site in Leith in 1820. It was a massive specimen, one of the largest elms in Britain, as it had grown unrestricted by competition throughout those nearly 200 years, resulting in a beautifully broad, round crown and open, graceful form. Its removal left a gaping hole in the garden. It could not be allowed to have died without due recognition.

The Wych Elm Project was born. The felled tree was debarked so that it could no longer host the disease-carrying beetles, and the bark was burned. Ian Edwards, who helped to run the project, said, 'I think we had all expected this huge tree would have a hollow core, but it was sound through to the centre.' It was taken to a sawmill at Cousland, where it was sawn into boards to be transformed into furniture and other handcrafted pieces by a selection of Scotland's top wood craftsmen and women, so that it could live on. These pieces were brought together in an exhibition at the Botanics. Meanwhile, community projects involved hundreds of people in building a wych elm yurt, making music, poetry and songs, and getting schoolchildren involved in woodwork. They made a film and Max Coleman created a book about it all, including not only profiles of all the makers and their elm pieces but also in-depth studies by the country's experts on the tree's evolution, ecology, folklore and human history. It's a beautiful book and will no doubt long remain Scotland's classic reference work on wych elm.

The many objects created from the Botanics' elm include a throne-like chair, a big engraved table, a chest for Ian Rankin's hallway, occasional and decorative furniture and also some gorgeous jewellery, abstract sculpture, toys, calligraphy and a fishing rod. There is even, defying convention, a harp, but not one for a chamber concert – Mark Norris made an aeolian harp, with thirty-nine metal strings, intended to be played by the wind, housed in a special pavilion – also made from elm – to protect it from rain but allow the breeze to whistle through. One of my favourite pieces is a bracelet, made from a single small elm piece by Geoff King, who carved it into a wrist-sized ring, one half of which is engraved into the form of an owl with outstretched wings, its feather patterns most beautifully following the grain of the wood. It seems appropriate for an owl to emerge from the tree in which it so often makes its home.

The lovely grain is commented on again and again by those who worked with the wood, which is described as 'a wild timber, with a grain pattern that is quite spectacular', by Chris Holmes. Michaela Huber created a stunning bookcase from her sections of the tree, showing off its green streaks like flames in the uprights that flank the spines of the books. The skill of a good woodworker is to give the impression that the tree lives on in its wood. This is elegantly achieved in a small table made by Ian Grant, which resembles a woodland animal, with an elm burr belly and antelope legs. According to the person who commissioned it, 'it looks like it has just walked into the room from the garden outside'.

The giant tree that had graced the garden since its inception certainly did not die in vain, and it lives on in all these vibrant works of art.[13]

Chapter Six

Life: Elms Around The World

The most delicate artwork created from the Edinburgh Botanics' elm was a pair of fairy slippers. Isabell Buenz made them from handmade paper, the pulp of which incorporated material from the elm. They remind me of the fairy houses constructed in the quiet corners of public gardens and tucked into hollows in the base of old trees – who knows by whom?! If a woodland isn't a place where playing children may encounter mysterious little creatures, how impoverished it would be! As a child I regularly came upon the signs of strange beings in the woods, and all around the world the folklore around elms is rich in tales of supernatural creatures and magical people, but these days forests, and especially tree plantations, are too often seen only as crops, or at best ecological laboratories, valued for their financial return or as 'natural capital' that will provide 'ecosystem services', quantified and priced and manipulated to serve the interests of our overprivileged species.

It hasn't always been like this, and it isn't like this everywhere. In the 1990s, when I was working as a forest activist,

Bill and I attended the United Nations' Forum on Forests for several years, alternately in New York and Geneva. Although it has attempted to tackle global deforestation and forest degradation, a consensus never emerged about any kind of international treaty – or, in UN jargon, a legally binding instrument – for protecting these most precious and threatened parts of the earth. Nonetheless, I found it fascinating as politicians, scientists, foresters, tree-huggers like me and, most importantly, representatives of Indigenous people whose homes are forests all came together. At the heart of these fora is the negotiation of UN policy, which is gone through literally word by word, sentence by sentence, paragraph by paragraph, giving any country the chance to amend it and seeking a consensus before moving on.

The day I remember so vividly was chaired by Sir Martin Holdgate, an eminent, schoolmasterly, bespectacled English scientist who banged his gavel with great authority each time a paragraph was agreed. For days he had been responding methodically to the requests for changes to the text by nodding to government representatives with a raised flag on their desk and 'recognising' their country, at which point they were able to politely make their request for a change of wording. Other countries would suggest alternatives or argue the text should remain unchanged, and only when nobody opposed could the debate progress. If such consensus proved impossible, the disputants went away to resolve the issue in private. We eventually reached a paragraph that described the values of forests to a wide range of different people and included the phrase 'small forest owners', a term popular in Finland and Sweden, where vast numbers of people own areas of commercial forest and Small Forest Owners Associations are powerful organisations.

The Indian delegation raised their flag to ask whether it was less ambiguous to refer to 'owners of small forests', on the basis that the phrase 'small forest owners' could mean it was the owners rather than the forests that were small. The European Union delegation, at frantic signalling from the Swedish government representative, promptly raised their flag and prepared for battle. Sir Martin leaned forward, smoothed his papers and launched into a charming speech about the ancient and strongly held beliefs in some rural, misty corners of Britain about diminutive people who inhabit woodland, often make mischief, dress in green and occasionally wear an acorn cap. Given that in other parts of the world, unknown to him, there may be other such people, he politely suggested therefore that there was merit in the ambiguity of the phrase and would the Indian delegate, whose grammatical clarifications had been so often helpful to the improvement of the quality and readability of the text, be willing in this case to live with it? The entire hall, usually humming with whispers and simultaneous translators, fell completely quiet. This eminent scientist had just advocated for pixies and leprechauns to be included in the official UN text! Nobody could believe their ears. The lead Indian delegate, with a wobble of his head, demurred. Sir Martin, with the merest twitch of a smile, stared hard at the Swedish and Finnish representatives, who silently lowered the EU flag. He tapped his gavel, small forest owners had their day, and the negotiations continued.

A Western scientific worldview is absolutely not the only way forests can be thought about, and thankfully other voices play a significant role in forest politics. UN texts acknowledge the social, cultural and spiritual value of forests, and the next few chapters of this book will explore these dimensions of elm trees. In my research, it has become clear that an Asian perspective is absolutely fundamental to our understanding of an elm tree as not just a body of wood, or a living being, but something that plays a crucial role in human culture and society.

Isabell Buenz's fairy slippers are made of paper, and it is this material that will guide us into the Asian story of elm, beginning with China. We often take paper for granted – many of you will be reading these words on it but probably weren't wondering what it is made of; from toilet tissue to our passports, from cardboard packaging to the great libraries of the world, paper is ubiquitous and fundamental to civilisation. The invention of this marvellous material is credited to Cai Lun, secretary to Emperor He, around AD 105, during the Eastern Han dynasty. According to legend, Cai Lun introduced to the emperor's court a new technology for the highly esteemed art form of calligraphy. A beautifully smooth-surfaced paper could be generated by mashing the bark of the qing tan tree, diluting it into a pulp, then dipping a gauze-covered frame, called a deckle, into the liquid, allowing the water to drain out, then peeling the resulting mat of fibres off the gauze and drying it. The best calligraphy paper, Xuan paper, is still handmade this way today, by a small family company in a town called Jingxian, in Anhui province. I visited it while researching my book *Paper Trails: From Trees to Trash – The True Cost of Paper*. There is, inevitably, dispute over the first paper – archaeologists in northern China found a letter written on flax paper from at least 170 years before Cai Lun's, during the

Western Han dynasty, and there are also claims that, 200 years even before that, in the third century BC, Khanzada people, from Tijara in the Alwar district of Rajasthan, India, first made paper similar to the lovely lokta paper you can buy in fair trade stores, probably using fibres from daphne plants. Regardless of which came first, what got me really excited is that Cai Lun used qing tan bark, as the makers of Xuan paper do to this day. When I enquired about the tree at the Hangzhou Botanical Garden, a curious and rather delighted curator took me to the *Ulmus* section of the garden. Qing tan is an elm!

Or, at least, it used to be. Unfortunately those pesky taxonomists I fell foul of in chapter two have gone and ruined the story. Qing tan's scientific name was formerly *Ulmus cavaleriei* but it is now classified as *Pteroceltis tatarinowii*. It looks like an elm, it smells like an elm, it pollards like an elm and it's found in botanic gardens with all the other elms, but it has in recent years been rudely expelled from the Ulmaceae family and shunted off into association with nettles, hemp and other fibrous plants in the Cannabaceae family. This is a great shame. Perhaps it will be rehabilitated into the elm gang one day, but until then, it's a nearly-elm with a fascinating story, which I simply had to include here.

The inner surface of elm bark is most useful. This is known as 'bast' and it consists of the phloem fibres, the tubes that transport the sugars made by leaves around the tree. As a key part of the living system of the tree, the phloem fibres are strong, flexible and waterproof, and in elms they retain these properties after they cease being used by the tree, which happens as the xylem and phloem are renewed each year. The previous year's phloem is squeezed flat against the bark so the bast consists of layers of really functional material. This fibrous inner bark of elms has had many uses across the world for many thousands of years. In my neck of the woods, as in many others, people made twine with it, which was used in tie-dyeing.[1] The string was wrapped around yarn or wool or cloth when it was dyed, to create patches where the dye would not take, which resulted in interesting patterns, sometimes reminiscent of the patterns of grain in wood. John Evelyn's seventeenth century treatise, *Sylva, or, A Discourse of Forest-Trees*, records the Royal Navy using ropes made out of wych elm bast. American elm bark fibres were woven into textiles by First Nations. Apparently it was still in use in Britain into the twentieth century for making matting used by plant nurseries. But only in China has the use of elm bark fibres been taken to its perfect extreme – by using it to make paper.

Calligraphy is arguably China's highest art form – encompassing the poetry of the words written and ideas they express, the oral music of their spoken form, the visual artistry of their presentation on paper and also, particularly in cursive styles of calligraphy, the physical dance of the calligraphic artist as they wield the traditional tools of paper, brush, ink and stone in the ceremony of writing. The nearly-elm is the very foundation of this art. The Xuan paper made from the qing tan tree is used not only for the finest calligraphy but also for some of the many other quintessentially Chinese artifacts that make paper much more than merely a functional material, rather one with spiritual value. Flying paper kites is not just fun – it is a way of making contact with air spirits or, at the very least, a kind of meditation. Masks made from paper are central to theatrical traditions for telling stories of gods, demons and mystics. Shrines are decorated with paper-cut art.[2] There is also a tradition of burning paper as a way of penetrating the veil between the living and the dead, and while some of the paper used for these 'ghost' transactions is cheap and poor quality, sometimes the finest quality paper is used to write prayers on, before setting it alight. Paper ghost money is burned as gifts to the spirits of ancestors, on special days like birthdays and anniversaries of deaths and on the national holiday dedicated for remembering them.

Given this deep cultural significance of paper in China, it's perhaps not surprising to find that there is a mythical creation tree, from the bark of which the first paper and the first people's clothes are supposed to have been made. Many cultures have a 'Tree of Life' and in China it is called Fusang, located on an island somewhere to the east, where the sun rises. According to one version, ten legendary ravens live in the tree, one of which carries the sun across the heavens every day. It was Hui Shen who wrote, in the *Book of Liang*, in the seventh century AD, about its bark being used to make paper, which links Fusang to the qing tan tree, and thus to elms.

China is not the only Asian country that gives legendary status to elms. There are also deeply revered elms in India, particularly the aptly named Indian elm, also known as jungle cork. Its Hindi name is *kanju* and it is called *chirabilva* in Sanskrit and also *udakirya*, which means 'tree growing close to water', which tells us that *Holoptelea integrifolia*, like most elms, likes moist conditions. It is widespread across India and into Myanmar. It's a useful tree, valued for its wood and also for its medicinal properties, with the juice of its leaves or a liquid made from its bark used as a treatment for intestinal worms – in humans and livestock – and also to reduce inflammation and tackle various allergies and viral infections, including flu.[3] It has been used as medicine for thousands of years, and its benefits and methods of preparation are documented in several Indian healing systems, including Ayurvedic medicine. As the world faces bugs that are untreatable by standard pharmaceutical antibiotics, medical researchers are looking towards traditional cures and indigenous wisdom. The stalwart elm of *chirabilva* is generating considerable interest as it seems to have a genuinely powerful mix of phytochemicals that are exciting medical scientists with their potential.[4]

From the damp to the dry, elms of one form or another thrive throughout Asia. *Ulmus pumila* – often called Siberian elm, though it is found from Turkestan to Korea – is one of the most tolerant of dryness, growing in some arid habitats, and there are magnificent specimens with girths of more than 5 metres in the Gobi Desert in Mongolia. Its fleshy samaras are stewed up and eaten and, just as in Europe, in hard times their leaves have been a famine food, such as during the Great Chinese Famine in the late 1950s and early 60s. In some parts of China, where trees are a vital defence against desertification, there are laws against cropping their leaves for food or fodder. The Siberian elm thrives in poor, well-drained soils and puts up with intense heat and deep cold, and because it grows very vigorously when young and is often resistant to Dutch elm disease, it has been widely planted around the world, including in Italy, where it was used to replace the field elm in viniculture; in Spain, where it was grown as an ornamental tree; and in North America, where it was planted as shelter belts and in hedges to help prevent the spread of the dust bowl. That fast growth and preference for poor soils has meant that it can become a liability, and it has spread widely in North America along railway lines and roadsides, becoming invasive in some areas, particularly New Mexico, where it is considered a real pest. It is one of the very few elms that is able to self-pollinate. It is short-lived in most places, barely outliving a human being, though in parts of its home territory, such as in the Russian Far East and northern China, some big, slower-growing trees have endured for hundreds of years.

In much of Russia and neighbouring areas, Siberian elm is called *karagatch*, though this name is also used further west in Turkestan and into Europe, where it is confusingly a name for a variety of field elm. The Russian explorer Nicholas Roerich, whose adventures were recorded by his son in a book *Trails to Inmost Asia*, found the welcome shade of a mighty double-trunked *karagatch*, 'lonely and towering' in the Mongolian desert – this one would almost certainly be the Siberian elm. 'The deep, deep shadow of the tree covered about 50 feet across. The powerful tree-stems were covered with fantastic burr growths. In the rich foliage, birds were singing and the beautiful branches were stretched out in all directions, as if wishing to give shelter to all pilgrims.'[5]

As we saw in chapter two, the elm family includes a group of trees called zelkovas. One of these (*Zelkova serrata*) grows in Japan and is known as *keyaki* in Japanese. It is used to make a big drum, called the *nagado daiko*, one of the special instruments for the traditional art of *taiko*, a musical performance that involves a group of players making a lot of noise with hollowed-out tree trunks, animal hide and big sticks. My friend Alison Roe performs and teaches *taiko* drumming, and she gave me a brief demonstration on her own *nagado* drum. *Nagado* means 'long body' and the drum is almost a metre tall, and fat as a Buddha, a big presence in the room. Alison is lithe and petite – she looks as if she would play a delicate little instrument, a flute perhaps. She approached the drum in a nonchalant manner, then broke into a dance. Her sticks flew and pounced and spun as she performed a kind of ballet, her entire body flowing with a complexity of movement at once effortless and profoundly formal, and out of this fluid motion came an extraordinary range of sound: a voice murmuring and tapping, clicking and hammering, booming and roaring. Her ceremonial dance was transformed by this big, shining, beautiful, round instrument into song. It was a brief but mesmerising display. I remember once seeing a *taiko* group at a festival and being blown away by the athleticism of the drummers, the complexity of their rhythms, the sheer noise of the massed drums and the ancient power of this ritual performance.

Alison said, 'Since, in Shinto, every natural object has a *kami* (spirit), a *taiko* therefore has a spirit, composed of the spirit of the tree its body came from and the spirit of the cow its skin came from, so when you play a *taiko*, you're playing with its spirit.' The making of *taiko* is an art form in its own right, and it can take four years from the felling of a zelkova tree to season the wood, hollow it out, shape and decorate it, carving the inside of the drum to give it its perfect sonic signature and then skinning it with leather.

Some of Japan's most important *keyakis* are not to be used for making *taiko* drums. One is the mind-bogglingly huge and ancient tree known as the Noma Keyaki, in the Arinashinomiya Shrine in northern Osaka Prefecture.[6] It is estimated to be a thousand years old, and at more than 14 metres, its girth is almost twice that of the biggest elm ever known in Europe. As a sign of its holiness, it wears a sacred rope like a belt strung with amulets around its trunk. It is a designated national monument, venerated by local people and visited by local farmers every spring, when the first leaves appear, to pray for a successful harvest. Although elms of this age are often hollowed out inside, this remarkably vigorous tree still seems to be fully intact, with a vast canopy and roots that cover an even larger area, one of which was found in digging works at a construction site several hundred metres away from the trunk. Even in winter it appears to have green leaves, but this is actually due to a kind of mistletoe, known in Japanese as *yadorigi*. Unlike the harmless epiphytic ferns and plants that grow on the elms near my home, this is a parasitic plant and it needs to be pruned regularly to prevent its theft of water and nutrients from damaging the tree.

The Noma Keyaki isn't the oldest in the country, however: that honour goes to a tree that may be up to 1,500 years old, and grows in the grounds of the Higashine Elementary School in Yamagata Prefecture. It is worth pausing to consider what this tree has seen over all that time as human empires have come and gone like passing flocks of birds.

Trees of that age are guaranteed to have stories to tell. There are many examples, such as another ancient zelkova, in the Wakamiya Hachimangu shrine in Hitachi Ota in Ibaraki

Prefecture, which is believed to have been the site of a visit by a famous Buddhist monk, Kobo Daishi, also known as Kukai, the eighth-century founder of Shingon, the Japanese branch of Vajrayana Buddhism. Legend also credits him with establishing the *kana* script, which is used for Japanese writing to this day. It is claimed – though like all good stories, it is also disputed – that he wrote the first *iroha* poem, a form that uses every character of the script exactly once, thus cementing what had up to that time been a more fluid set of symbols. The story goes that a scruffy, poor-looking monk once came begging at a house that was built in the shade of this *keyaki* tree, but the housekeeper sent him packing. When the owner of the house discovered that this was in fact the great holy man, he climbed up into the top of the tree and shouted apologies, calling to him to return, promising honourable treatment. Eventually he was magically turned into a cicada, and ever since then, every year on 23 July, the day when Kobo Daishi's visit happened, cicadas arrive and begin singing in the old zelkova.

Around Japan there are many other stories linked to *keyaki* trees, particularly involving the presence of a *tengu* (spirit creature) that lives in or under the tree. Some of these are benign, like the one who visits the local sake-maker's workshop to assess the brew, sipping noisily if it is good but only quietly if it is less so. Others are tricksters, who steal children away, rather like the fairies of Celtic legends, returning them later with unlikely tales or excuses for their absence, or in the worst cases, not returning them at all, but eating them! There are stories of *keyaki* trees that were cut down, causing the death of the person or people who felled them. There's a *keyaki* that will cure warts. And, of course, there are *keyaki* haiku, like this one by Tada Chimako (translated by Hiroaki Sato).[7]

葉を脱いで欅すらりと月の中
ha o nuide keyaki surarito tsuki no naka

> having shed its leaves
> the zelkova stands svelte
> in the moonlight

In the north of Japan and the Russian islands around the Sea of Okhotsk, the Indigenous Ainu people worship a goddess called Kamuy-huci or Kamui Fuchi), who is believed to be the daughter of a Japanese elm tree (*Ulmus davidiana*), fathered by the original creator of the world and god of the sky, Kandakoro-kamuy.[8] She is one of the most powerful Ainu goddesses, and she governs the gateway between the human world and the world of the spirits and gods. In Ainu cosmology the hearth embodies this portal, with the spirits of the dead being conceived of as beneath the household fire, and the hearth is therefore profoundly sacred. From her place in the hearth, Kamuy-huci watches over all of domestic life, a kind of spiritual housekeeper in charge of various lesser spirits, including one looking after the garden, and Rukoro-kamuy, the goddess of the privy. Ainu people traditionally used fibrous elm bark for many domestic purposes, including making cloth and thatching their houses, and so it makes sense that the hearth goddess would have an elm tree for her mother. Kamuy-huci lives on in contemporary Japan through Typhlosion, one of the creatures in the Pokémon game. The Japanese elm, also known as Father David's elm, is grown in many other countries now as it has strong resistance to Dutch elm disease.

Even more widespread is another Asian species of elm, *Ulmus parvifolia*, which is universally known as Chinese elm, although there are several other elms in the country and it isn't restricted to China anyway. It is one of the most suitable species for the curious tradition of bonsai, the process of growing a tree in a very small pot for a very long time, so that it achieves the optical illusion of being a huge and ancient landscape feature when placed among rocks in even the most diminutive garden. Chinese elms respond to the harsh treatment of root-stunting by contorting, as if raked by storms, thus taking on the appearance of an ancient tree within relatively few years.

However, if a Chinese elm is allowed to grow freely in good soil, it is an attractive and extremely hardy tree, surviving at temperatures as low as −34°C. It produces hard wood that is excellent for making tools, as is true for most elms, and it hardly ever succumbs to Dutch elm disease.

Elm trees are not native to the Southern Hemisphere. In the Americas, they get as far south as Costa Rica, but that still puts them 10 degrees north of the equator. *Ulmus glabra* (wych elm) is reportedly found growing in the wild in North Africa, although it is thought by Eddie Kemp, the former curator of Edinburgh's and Dundee's botanical gardens, to be introduced rather than native.[9] *Ulmus minor* (smooth-leafed or field elm) is the most common elm in Southern Europe and Asia Minor, and though most accounts state that its natural distribution doesn't take it further south than Iran, it is also recorded as a native species in Israel, Tunisia and Lebanon.[10] Even so, these are still firmly Northern Hemisphere.

South of the equator, it's the mysterious wych and field elms in Africa which have some botanists shaking their heads and declaiming they aren't native. This hotly debated topic occasionally results in acrimonious arguments about 'aliens' and 'invasiveness', dragging governments into some very expensive eradication programmes.[11] The question of where elms actually belong raises questions about balancing the risks of planting exotic species and the potential benefits of creating ecological refuges.

But first, there are two interesting trees to meet. The African elm isn't an elm at all, or at least not anymore. Another victim of taxonomists, *Celtis africana* was chucked out of the elm tribe in the same manner as the qing tan tree and it is now part of the Cannabaceae family. Still, it looks like a small elm and is called an elm by many English-speaking people in the various African countries where it grows from Yemen through Kenya to Tanzania and further south to Zimbabwe, Mozambique, Botswana, Lesotho and South Africa. It's also known as white stinkwood. There's also a tree sometimes known as thorny elm that grows in northern South Africa, but that's not ever been recognised by botanists as an elm and is, scientifically, *Chaetacme aristata*, and completely unrelated except in its common name.

The one and only – as far as I'm aware – true member of the elm family to have ventured south of the equator under its own steam is *Ulmus lanceifolia*, which as its name suggests has spear- or lance-shaped leaves, though these are maximum

10 centimetres or so. The tree can grow to be huge, up to 45 metres tall, and it is genuinely tropical, growing across a broad swathe of South-East Asia, from Eastern Himalayan India all the way through Bangladesh, Myanmar, Thailand and on to Indonesia, where it tips over the equator. It is also found in southern China, Laos and Vietnam, where it gets its English common name of the Vietnam elm. Given that it is happy in the forests of Sulawesi and Sumatra in Indonesia, it's not surprising that it doesn't easily survive European or American winters and is rarely seen in arboreta or botanical gardens, although there is one among an impressive collection of elms at Grange Farm Arboretum in Lincolnshire, England.[12]

Elms are found in considerable numbers in several countries in the Southern Hemisphere, however, because people love them and want them to line their streets or grace their gardens. The aesthetic value that led to tens of millions of them being planted in pretty much every city in America, their practical benefits of shading and shelter, and their lack of antisocial habits like dropping hard seeds on people's cars or growing roots into drains have made them a favourite with urban planners throughout the world. Plus there are so many different varieties of elm that inhabit such a range of different habitats that, wherever you are, there is likely an elm that will enjoy living in your garden. In South Africa, for example, the Chinese elm is recommended for gardeners as 'a handsome shade tree'.[13] There are also plenty of field elms in one or other of their colourful variations on sale in South African nurseries. But to see elms as a significant part of urban and rural landscapes in the Southern Hemisphere, we need to go to the former British colonies, where people emigrating from England, in particular, took with them the seeds and scions of the elm trees they valued so much. Australia and New Zealand have loads of elms, both as a result of two centuries of British-influenced landscape gardening, and also due to global isolation making them far less prone to the Dutch elm disease that has decimated elm populations in the Northern Hemisphere.

When British people colonised Australia, fast-growing elm trees were a potent way of making the unfamiliar landscape resemble their homes. Visiting Melbourne in my early twenties, I was taken aback to be in a city that felt like somewhere in continental Europe. I had emigrated to Australia in 1989, aged twenty-three, having scored an academic job at Griffith University in Brisbane. Many of my expectations about life and work down under were dashed immediately on arrival, and the others were gradually eroded within a year. The gum trees seemed skinnier and greyer than I'd pictured, and I had to beware of spiders that could kill me. I had expected that by moving to the 'new world', I would escape the sexism and ageism of British academia. Instead, Queensland society was even more reactionary. In my first few weeks, one of my fellow lecturers casually asked me to do his photocopying for him, and when I refused he took umbrage on the basis that as a young woman I would be 'better at it than him'. I was also horrified to discover that not only were there no Aboriginal members of staff, there were not even any Aboriginal students. The casual racism was appalling. Homosexuality was illegal and I took to the streets to join joyfully in the massive street party when it was decriminalised.

I loved the forests, was utterly bored by the climate, with its endless good weather, and quickly decided that I was completely unsuited to the country. One too many pub arguments about racism ended with my antagonist saying, 'If you don't like it here, Pom, go home.' I handed in my resignation, planning a circuit of the country by campervan before I left. In the course of that journey I discovered that there were many different Australias, and as I had only experienced Queensland, I had a distorted view of the country as a whole. Melbourne was one place that redeemed the nation in my eyes – politically liberal, with complicated, changeable weather and oddly familiar green vegetation, including elm trees.

Melbourne is home to thousands of elms, mostly English and Dutch (*Ulmus minor* 'Atinia' and *Ulmus* × *hollandica*), some of which date back to the early days of British colonisation. Philip King, the third governor of New South Wales, imported elms from England in 1803, and some of those early nineteenth-century trees may survive. One of a group that was planted in 1846 as part of the establishment of Victoria's Royal Botanic Gardens is still standing.

Melbourne boasts one of the most gorgeous displays of urban elm grandeur, in the form of the tree-lined avenues of Fitzroy Gardens, designed in 1859 by Clement Hodgkinson, planted in the 1860s and tended in subsequent decades by head gardener James Sinclair. For the first few decades, the park was reportedly pretty unpleasant, with a lot of swampy ground and a tributary of the Yarra River running through it, which acted as an open sewer. Elms were densely planted along with many other trees, both native gums and exotic conifers, with lots of willows and ferns taking advantage of the wet conditions. By the end of the nineteenth century there were improvements in sanitation and hence water quality – and smell – and the woodland was thinned out; this included removing every other elm along the avenues, giving the remaining elms space to expand as they matured. The result is that the walks have the lovely tunnel effect of trees interlocking overhead, providing dappled shade and the welcoming come-hither sense of a path stretching off under the shelter of foliage.

Handsome lines of elms occur in plenty of other places in southern Australia, especially in the state of Victoria. They became fashionable in the late nineteenth century, and then after the First World War there was a spate of planting avenues of honour, and some of these are exclusively or primarily elms. The elm is often used to commemorate lost lives; in Camperdown, Victoria, a tree was planted along a roadside for each local soldier killed in the war. In some cases the carnage was so great that they are many kilometres long, such as the 22-kilometre avenue in Ballarat. Other elm-dominated avenues of honour are found in Digby, Bacchus Marsh, Creswick, Newstead, Traralgon and Wallan.

So far, Australia is completely free of Dutch elm disease, and it is a place of refuge for many different species planted around the country. Melburnians can claim one of the greatest concentration of European elms in the world, with up to 6,000 in the centre of the city and at least a further 10,000 within a 10-kilometre radius. The state of Victoria records more than 30,000 elms on state land with at least as many again on private land. So, for non-native trees, Australia's elms are certainly numerous and they are a globally significant population of a range of elm species, although one that is at high risk from Dutch elm disease should it arrive, having had no historical exposure. It is little wonder that Australia is notoriously strict about plant-based imports.

This beautiful refuge for elms is not without its cost. In Fitzroy Gardens, a scarred tree shows the use of its bark by the Aboriginal Wurundjeri people, whose lands were taken by British colonists. The Yarra River was theirs, and never as polluted as it became after colonisation. Most of the native trees that would have grown where Melbourne now stands are long gone. Those beautiful parks with their lovely elm avenues flourish in the soil of historical injustice. The elms in Australia are invaders, yet they have rooted deeply, like many of the people who settled there. They are diverse in species as Australians are diverse in ethnicity themselves. What priority should we give Indigenous species and people? All I knew was that I felt deeply like I didn't belong, so I went back to where I came from.

Across the Tasman Sea in Aotearoa, the story of elms is entirely different, though it begins similarly. When British people and other Europeans started colonising the islands they called New Zealand, they brought elm trees, primarily to plant in their gardens and help them feel at home. When I visited in 2017 I was staggered by how much land was taken up with monoculture conifer plantations, as well as by irrigated cattle pasture. Elms, as far as I'm aware, have not been widely planted by commercial forestry for their timber, unlike pines, but even so they are grown widely for their shade, shelter from the elements and as an elegant backdrop.

A big house and gardens called The Elms at Te Papa Tauranga on the east coast of New Zealand's North Island, south of Auckland, is a historic site for Māori people. It is long inhabited, an important landing spot for *waka taua* (war canoes) and the location of much intertribal battling. As the number of non-Māori people living there began to increase in the 1820s, the missionaries took advantage of post-conflict devastation to establish a mission station, which included a garden with an abundance of elm trees from Europe. The official history of The Elms claims that 'local chiefs realised the advantages that a missionary presence could provide for trade and security'.[14] Led by Alfred Brown, it began operating in 1838 but in the 1860s was made headquarters for British imperial troops in their battles for dominance over the Māori. After resistance had been squashed, in 1873 Brown bought the property privately, by which time the elm trees had become sufficiently well grown that he named the house after them. It has been The Elms ever since, and it stayed in the Brown family until 1999, when it became a charitable foundation.

According to a 2016 study by plant pathologist Rebecca Ganley, over 400,000 elms have been planted throughout New Zealand.[15] About 30,000 of these are in the major port city of Auckland, where their environmental benefits, aesthetic appeal and positive impact on real estate value are estimated to be worth NZ$200 million. Across the country, with the conservative estimate of NZ$1,000 per tree, that value can be at least doubled. As we saw in chapter three, the loss of city trees when disease strikes is extremely expensive.

It was therefore with horror that the country woke up just before Christmas in 1989 to hear that a tree in an Auckland park was dying of Dutch elm disease. We will never know how it arrived, but the smaller European elm bark beetle (*Scolytus multistriatus*) was spreading the fungus and it was assumed that beetle grubs had hatched out from infested elm packing material in a container ship. Most of New Zealand's elms are European species with little or no resistance to the blight, and as in other countries, without action, tree loss would be at least 90 per cent. A national eradication programme and a clamp down on imports and movements of any elm material followed. Ganley said, 'The programme followed a strategy of locating all elms in the infected area, inspecting all known elms at least three times during the flight season of the vector beetle, and felling, fumigating and burying all infected material immediately after detection.' They also set pheromone traps for the beetles and, in short, went all out to try to stop the blight dead in its tracks.

They so very nearly succeeded. At first the disease spread fast throughout Auckland, with wilting trees spotted in an ever-widening radius as the beetles flew during the long summer months. After NZ$4 million was spent on the eradication attempt, gradually the incidents started to dwindle. In 1993, there was an outbreak further south, in Napier, which was stamped out completely, raising confidence that the disease could be beaten. In 1997–8 and 1998–9 there was only one case each summer, and the latter was a carry-over from the previous year, with no fungus-carrying beetles detected. At the turn of the millennium, plant pathologist Peter Gadgil and colleagues writing a chapter about New Zealand in a global survey of elms, concluded that 'eradication of DED from New Zealand is possible'.[16] There was a collective sigh of relief and in 2004 official notification that the disease control process would be stepped down.

Unfortunately that was not the end. In 2008 cases were continuing to crop up, leading to formal recognition that eradication was not in fact possible. Instead, the country attempted to slow the spread of the disease. Although it had been highly contained to the central Auckland area until 2012, that summer it spread from the heart of the city. Still, efforts to limit the beetles' opportunities to breed were effective, and by 2016, plant pathologists were confident that there were no cases south of the Greater Auckland area.

However, cases moved steadily south in the intervening years, and beetle traps were laid following the first confirmed case in North Waikato in 2021. In February 2024, there was a new outbreak 150 kilometres further south, in the agricultural town of Te Awamutu in the Waikato region, where hundreds of elms grow on council land, and many more on private property. Greater Auckland had clearly failed to curb the blight. When an elm on private land in Te Awamutu was diagnosed with Dutch elm disease, James Richardson, arborist for Waipā District Council, sought out other cases, performed sanitation felling and urged private landowners to take the disease seriously and dispose of diseased trees carefully. 'We're doing absolutely everything we can to find affected trees and limit the spread of the disease,' he said.[17] He believes that climate change is exacerbating the risk from Dutch elm disease. 'Probably the longer periods of warmer days have affected how active the beetles are, which are the main vector of the disease . . . They are more active for a bit longer each year.' And, as a result, sadly, they seem to be moving inexorably further and further across the country.

The Troll Tree

About 15 miles inland from where I live is Loch Assynt, a long thin loch that lies between the mountains of Canisp to the south and Glas Bheinn and Quinag to the north. The loch is fed by various rivers, one of which, on the north-eastern tip of the loch, is the Traligill. Its name comes from the Norse and means 'glen of the trolls'. The river has carved an impressively steep gorge, but if you battle upstream from the loch there is a place where the water bubbles up out of the ground, from a *fuaran*, the Gaelic word for spring. Beyond this point, the river valley is dry, the rocks exposed. Further along, the river runs again, but the water falls over a rock and simply vanishes underground. Between the disappearance of the river and its re-emergence is like a desert river valley, clearly carved by water, with rounded stones in the bottom and steep sides, but no water running. Yet here there are elm trees, one of which is huge, with a magnificent trunk festooned with mosses, lichens, polypody ferns and fungi, a rich tapestry of rainforest life. Uniquely, it grows horizontally out of the rock, many metres up the sheer wall of the ravine, a completely implausible place for a tree to grow, hanging in perfect defiance of the laws of physics.

I stand beneath it, neck craned in awe, looking up into the lush green profusion of its living community. It is winter, so all this greenery isn't the tree's own leaves, but photosynthesising life using it as a climbing frame. Paradoxically, in this dry river valley, everything about its grand gathering of epiphytes declares it to be a rainforest tree. It is a perfect symbol of survival against the odds.

As I marvel, I realise that I can hear mumbling, murmuring and muttering. Behind the rock from which this great trunk is buttressed, trolls are grumbling. There are distinct thumps of their hammers and the sharp percussion of stone being worked, irregular hissing and, underneath it, deep groans and roaring. They sound busy and bad-tempered, these subterranean beings.

Looking back up to the tree it has become even more magical – if there are trolls below, is this an ent? It's not difficult to imagine it leaping down from its rocky perch and striding off down the glen. Is it in league with the trolls or keeping them under control? How does it survive in this arid zone, halfway up a crag?

There is, of course, an explanation for this mystery. The Traligill River emerges out of an area of limestone riddled with caves. This part of Assynt is famous among geologists and speleologists (cave explorers) for its network of caverns and underground channels. This underground world is a treasure trove for animal historians and zooarchaeologists because of the Bone Caves where the remains of animals – including polar bears, reindeer, lynx and the UK's most recent brown bear bones – have been found. Beneath the magnificent elm, the Traligill River is still flowing, and rumbling invisibly beneath our feet.

The tree's roots are presumably taking advantage of the moist dark network of cracks and crannies to supply it with all the moisture and support that it needs. It hangs well above the riverbed because with ice-melt or torrential rain the river can burst out from the limestone cave system below and flow ferociously. Up there on the crag, the elm is at much less risk of being swept away.

When I first moved to Assynt, twenty-five years ago, I weathered frequent and seemingly inexplicable emotional storms. There are many things about living in this gentle, friendly, argumentative, beautiful, wind-ravaged, sea-buffeted and biodiverse corner of the world that have helped me over the ensuing years, but from early on I realised that it was the trees which had the most to teach me. This troll tree is the perfect metaphor for my own psyche. All of the tree's invisible matter is akin to the aspects of ourselves that are never shown to the outside world, often not even registered by our conscious minds. Like many writers, I have experienced the magical way a story or a poem emerges seemingly of its own accord from somewhere other than consciousness, appearing on a page via a hand that is watched with amazement by an eye that can't understand where it has come from. A character walks into a story entirely of her own volition. Words rhyme and chime and sing without any conscious effort. The way that such creativity flows reminds me of how a tree's above-ground flourishing is invisibly fed and nurtured from underground or within a rock. As I try to heal from the life and death of my father, it is the subconscious part of me that is doing a lot of the important work. Metaphors of all kinds have helped me in this struggle about who I am. The elm most of all.

Perhaps the elm's most important lesson is wintering. All of our deciduous trees cast aside their leaves in autumn and sink into dormancy. They rest for months on end. During this time they are battered by storms, ice, snow, hail, rain and more storms. They withstand it all, perhaps pruned but largely unscathed. What they are up to underground is anyone's guess, but mostly they rest, while life goes on around them.

The elms have taught me to rest too, and meanwhile I wait for the elm trees to come into flower. As their buds start to show, up high in the silhouettes of their twigs, bulges form, promising something new.

I return to the troll tree in spring to gather seeds. I'm sceptical that the walk up the glen will be fruitful, but it's a lovely day with patchy blue sky and enough breeze to keep midgies at bay. At this time of year there are always flowers growing on the sweet limestone-based soils that I don't see on our peaty acidic ground closer to home. Right on cue I stumble over a miniature plant with a stalk full of red globules tipped by tiny white flowers. I later find it in Mum's flower book: it's alpine bistort and it gets two stars, which means it's pretty rare. Just as exciting is the discovery of the ravishingly beautiful globe flower, with leaves like a buttercup and a flower which is exactly what it says: a perfect sphere of delicate yellow petals. It is growing just outside the span of the elm, which is in its full summer glory, bedecked with leaves, from the bottom of the dry river valley all the way up to the top of the ravine. It is the tallest tree around by far. Its trunk emerges halfway up the crag and its canopy spreads vertically from the ground to the sky, as if someone, a troll presumably, has turned it on its side, like an umbrella left to dry. The trolls,

I'm pleased to note, are rumbling peacefully down in their cave below. And I'm even more delighted that the tree is festooned with fruit, as are several other, smaller trees in the gully, and some of the papery samaras I pick have the fattened centres that indicate the presence of seeds. I gather up a bagful, telling the trees I'll give them to the tree nursery to germinate and grow. Their offspring will help our community to restore the rainforest of which they're such a ravishing part.

Chapter Seven

Death: European Elms and their Folklore

There are three elm species native to Europe: *Ulmus laevis*, *Ulmus minor* and *Ulmus glabra*. *Ulmus laevis* is usually called the European white elm due to the light colour of its wood, or, more beautifully, fluttering elm because its flowers and fruits are on longer stalks than most other elms, so they catch spring breezes. In North America it is known as Russian elm, and its range does cover a large part of Russia, from beyond the Ural Mountains in Asia all the way west. Its European coverage includes Spain and France in the west, north to Finland and south to Bulgaria. It is very much a riverside tree and loves having its feet wet, tolerating significant flooding and suffering in drought.

Ulmus minor is the field elm, occasionally called red elm due to its deeper-coloured timber. The field elm has been considered so useful around Europe that people have greatly modified its natural distribution over the past 10,000 years or so. Without human interference it is the most southerly of the European elms, stretching all the way to Iran and over into North Africa. It might have reached only to the Baltic naturally, but its range has been greatly extended northwards through deliberate planting. It is another lover of riverine habitats. It has been a highly contested species: some taxonomies used to distinguish a whole range of different elms including English elm, small-leaved elm, Plot elm, Cornish elm, Wheatley elm, narrow-leaved elm and various others, while other botanists, notably Richens, argued that they were actually all the same species. DNA analysis is clarifying that all these different elms are indeed just different varieties of *Ulmus minor* or hybrids between it and another elm. Some of them turn out to be genetically identical clones, including English elm, which originated in Italy and was probably brought to Britain by the Romans.[1] English elms are sterile, never setting seeds, propagating instead by underground shoots called suckers. Most elms avoid pollinating themselves, and when English elms are surrounded only by clones, the pollen from their neighbours is identical to their own, so they won't use it; even when they have other neighbours with which they could cross-pollinate, they don't. Their lack of genetic variability is one of the reasons they are all very susceptible to Dutch elm disease.

Ulmus glabra is Europe's third elm, commonly called wych elm. It also gets the moniker Scots elm, which is surprising given that its range is vast, stretching from Ireland in the west, deep into Russia in the east, from right up in northern Norway, far into the Arctic Circle and south to the Mediterranean Sea. Presumably Scots elm was a name to distinguish it in the UK from all the field elms – and certainly the further north in Britain, the less likely any other species becomes. It is, in fact, Britain's only native elm. The further south it grows, the more it prefers high altitudes, so it's often thought of as a montane tree, and is sometimes called mountain elm, and even in the north it will grow right up to the tree line, where it gets cool temperatures and high humidity. Its preference for moisture means that it is often found in stream gullies and on riverbanks, just like the other elms. It is a true temperate rainforest tree.

Right across Europe, Dutch elm disease has been so severe over the course of the last hundred years that many people's sole experience of elms is when they were younger, before watching them die. Many associate elms with death, especially older people who have seen this transformation. Younger folk often have no knowledge of elms at all, but historically they were a key part of many people's sense of place, their shapely forms acting as a backdrop to rural life and urban experiences. Colin Tudge, for example, says, 'Elms until recent times were so common in England they largely defined the lowland landscape: they dominate Constable's Suffolk landscapes in the east and in the west were known as the "Wiltshire weed".'[2] Their removal changed people's views of the landscape around them. Elms were the guardians of the hedge, and their loss was followed by the loss of many hedgerows and their associated biodiversity from English agricultural land over the past decades, as the prickly hawthorn that had nurtured these trees was now 'just a green thing that stands in the way', to quote William Blake. Over recent decades elms have come to represent environmental loss but their connection with death goes much further back.

The tale of the Greek hero Orpheus, the wondrous lyre player, unfolds on the death of his beloved wife, Eurydice. Orpheus follows her to the underworld, the land of the dead, unable to bear being without her. He is able to enchant Hades and Persephone, the god and goddess of the underworld, with his lyre-playing, and they agree to allow Eurydice to return to the living world, but she must walk behind Orpheus and he must not look at her while they make the arduous journey back to life. Of course, the inevitable happens and he looks back, trapping her. Why he turns is a detail that varies: either she stumbles and he turns to help her, or her footfall becomes so quiet he can no longer hear her behind him, or he loses faith that she will make her way out without his help. According to the earliest versions of the legend, having lost Eurydice, Orpheus stopped to play her a love song, and the natural world responded in the most extraordinary way to this heartbroken music: with elm trees.

The first elm grove was thus conjured from a love so passionate it transcended death, and through music so beautiful that it could entrance even the most powerful gatekeepers of the underworld. The elm, deeply rooted in the subsoil, has ever since stood as a symbol of the ambivalence and mystery of the end of our lives, wherever we go afterwards, and the power of love and art to help the dead live on in our hearts and in nature.

Elm's role as guarding the entrances to the underworld and having a role in protecting the dead might draw on elm's use for bows and coffins, or perhaps it is simply because it is often planted as an elegant shade-tree at grave sites. Whatever the reason, elms stand mythologically on the border between life and death. Richens says, 'The traditional date for felling elm for timber utilization was All Saints' Day (1 November).'[3] Although it is chosen for practical reasons – winter dormancy has begun – the morning after Hallowe'en is when the veil between the living and the dead is believed to be thinnest.

In some European countries, including England, an elm stake was driven through the body of someone who took their own life. Suicide has long been misunderstood by the Christian religion, considering it to be a sin of such a wicked form that people who have died by suicide cannot be buried in consecrated ground, with extreme methods used to ensure that their spirits do not contaminate the living. Why an elm stake should serve this purpose is anyone's guess.

In Cheltenham, a local legend from around 1700 tells of Maud Bowen, a poor but beautiful girl who flung herself into a stream and drowned while trying to escape from the sexual advances of the lord of the manor. Aided by Maud's wicked uncle, who was shot by a mystery archer, the landowner protested his innocence, which led to a coroner's verdict of suicide. She was buried at the road junction close to where she had drowned. An elm stake was put through her innocent heart and took root – elm will grow from cuttings, so this isn't completely unreasonable – and it grew rapidly into an elegant and beautiful tree.

The lord of the manor found Maud's mother, Margaret, grieving under the tree and deemed her to be in the way of his carriage. When his remonstrating servant was also mysteriously shot by an unknown archer, Margaret was condemned as a witch and burned at the stake close to the tree. During this grisly ceremony, the lord of the manor himself was shot by the archer, who is revealed at the end of the legend as a man who loved Maud, seeking vengeance for her and her mother's unfair treatment.

Called Maud's Elm, it was a vast tree, eventually more than 6 metres in girth, but it sadly succumbed to Dutch elm disease in the twentieth century.[4]

Elms have featured in Western European storytelling for as long as we can remember, playing an important part in *The Iliad* by Homer, whose poetry was oral – sung, in fact – and not written down. *The Iliad* tells of the siege of Troy, where the Trojan prince Paris has taken his beloved Helen, the most beautiful woman in the world, who is inconveniently married to the Spartan king Menelaus. *The Iliad* only gives us a slice of the conflict, ending far before the notorious Trojan horse, but it is the granddaddy of all epic stories and provides us with fascinating insights into the Mediterranean world of almost 3,000 years ago.

There are two direct references to elm in the poem. Achilles, the story's hero, is insane with grief at the death of his beloved Patroclus and goes into a killing frenzy. The River Scamander, which is the physical embodiment of the river god Xanthos, becomes enraged at the bloodbath that has left so many corpses in his water. Here is Samuel Butler's classic translation:

> The river raised a high wave and attacked [Achilles]. He swelled his stream into a torrent, and swept away the many

dead whom Achilles had slain and left within his waters. These he cast out on to the land, bellowing like a bull the while, but the living he saved alive, hiding them in his mighty eddies. The great and terrible wave gathered about Achilles, falling upon him and beating on his shield, so that he could not keep his feet; he caught hold of a great elm-tree, but it came up by the roots, and tore away the bank, damming the stream with its thick branches and bridging it all across; whereby Achilles struggled out of the stream, and fled full speed over the plain, for he was afraid.[5]

I recognised my own experience in this scene, awash with the psychological struggles brought up by my father's death, reliving and battling with childhood trauma that I had ignored for so long, having my own emotional bloodbath and feeling like I could not keep my feet either. And, like Achilles, I have been protected by the elm. In this scene, the great tree, though uprooted, dams the flood, calms the torrenting waters and provides our hero with a means of escape from the horror. Healing from trauma can include an intensely physical experience of needing to run, fast, to flee to a place of safety. I have found it strange but hugely helpful to allow my body and imagination to flee 'full speed over the plain' when the horrible memories erupt and I am afraid.

Homer's second elm reference occurs when Andromache, the wife of Hector, who is the great hope of the Trojan army, tries to persuade him not to fight Achilles. She tells how Achilles killed her father, a warrior called Eëtion. Oddly, her report reflects on Achilles as an honourable 'goodly' man, because having killed Eëtion he dealt with the corpse with great respect – 'he despoiled him not, for his soul had awe of that' – cremating the body in his full armour and then building a tomb, or barrow, over him, 'and all about were elm trees planted by nymphs of the mountain, daughters of Zeus'.[6] Why should the nymphs of the mountain plant elms around his grave?

No reason is given in the poem, but we can put two and two together. Eëtion was the initial captor of Chryseis, whose struggle for freedom is the inciting incident of the whole story and who, throughout, is treated as a mere chattel and sex-slave. So perhaps the mountain nymphs planted elm trees on his grave in sympathy for the woman who was abused by him. The elms reflect a female perspective, the story seems to say. The Greeks were not the only culture to make strong links between elms and women.

There is a third indirect elm in connection with *The Iliad*, which is that the first Greek to be killed in the Trojan war was Protesilaus, king of Pteleos in Thessaly, which takes its names from the Greek word for elms (*ptelea*) due to their abundance in that region. Elms were planted on his grave too, but when they grew tall enough that their upper leaves could see the city of Troy, they withered away. In the first century AD, poet Antiphilus of Byzantium wrote:

> Thessalian Protesilaus, a long age shall sing your praises,
> First of the many destined to die at Troy;
> The nymphs, across the water from hated Ilion
> Covered your tomb with thick-foliaged elms.
> So great was the bitterness of the heroes,

> It lingers, hostile, in their soulless upper branches.
> Trees full of anger; whenever they see that wall
> Of Troy, the leaves in their upper crown wither and fall.[7]

It's tempting to see in this story a reference to folk knowledge of some kind of blight, perhaps even the earlier wave of disease, as posited by palaeobotanists to explain the Neolithic elm decline.

I work as a literature and creative writing lecturer and teach a module which looks at epics from the earliest classics to the present day. After covering *The Iliad*, we move on to the story of the founding of Rome by an exile from Troy, Aeneas, as written by the Latin poet Virgil in *The Aeneid*, several centuries later. Epics throughout the ages all feature a flawed hero emerging from a world in crisis, going on some sort of physical or metaphorical journey, meeting challenges, overcoming them and bringing back some kind of talisman that can overcome or heal his world. Aeneas wants his dead father's blessing to found Rome, and to get it, like Orpheus, he must travel to the underworld, where he finds furies, harpies, gorgons and other legendarily dangerous monsters representing different causes of death. Among these horrors, astoundingly, is an elm tree. In W. F. Jackson Knight's translation, two short sentences present it to us: 'In the centre is a giant and shady elm-tree, spreading branches like arms, full of years. False Dreams, so it is often said, take the tree for their home and cling everywhere beneath its leaves.'[8]

It is no accident that the elm is among all these possible reasons why someone might end up down in the underworld, though it provides a strangely peaceful interlude, with its sad, clinging community of false dreams, between the violence of war and the monsters of natural disasters. It is as if Virgil wants us to acknowledge that by no means are all deaths the result of violence or horror: many are losses caused by hopelessness, and it is these who the elm tree cradles in its arms. Matt Haig was surely thinking of this when he cast Mrs Elm in *The Midnight Library*, as the librarian of all those books of dreams and regrets in the world between life and death.

While the elm as a classical symbol of death is pervasive, it is matched by another equally widespread and happier Roman symbol: the elm as a man entwined with a woman in the form of a vine. The traditional use of elm as a support for grapevines in vineyards was widespread in Italy when the great poets of Latin were writing. Catullus wrote in *Carmina* about a vine, 'If haply conjoined the same with elm as a husband, / Tends her many a hind and tends her many a herdsman,'[9] presumably implying great fruit production when trained up an elm tree. Cato, in the oldest known book of prose in Latin, *De Agri Cultura*, from the second century BC, wrote about pruning trees for the training of vines: 'The trees are to be pruned thus: the branches that you leave to be well separated; cut straight; do not leave too many. Vines should have good knots on each tree-branch.'[10] The Roman agronomist Columella wrote about the elm tree training method in his agriculture handbook *De Re Rustica*, in the first century AD, recommending the use of an infertile Atinian elm, better known now as English elm. Columella had vineyards in both Italy and his Spanish home of Cádiz, and he took the Atinian elm to Spain to grow his vines up it, as it was so perfect for the job, being tolerant of poor soils and of heavy pruning, the result of which was an abundant supply of leaf hay, preferred by cattle over any other tree prunings. The combination of its use in vineyards plus its value as a fodder crop led to it being dubbed 'the tree of milk and wine'.[11]

The widespread practice of vine-training up elms led to elm-and-vine becoming a metaphor for marriage throughout the Renaissance and the Neoclassical and Romantic periods, and it lives on into the present day, for example in the publisher Elsevier's logo.[12] Ovid provides the archetype in his poem *Metamorphoses* about a beautiful young woman, Pomona, a fanatical gardener who resists the advances of many suitors, including the god Vertumnus, who disguises himself to attract her attention and persuade her to marry him.[13] In the form of an old woman, he points out the benefit that her fruit plants are getting from their supports, and that she would gain similarly from the support of the handsome and good Vertumnus:

> There was an elm tree opposite, a lovely sight to see, with its bunches of shining grapes, and this the god praised, and its companion vine no less. 'But,' he said, 'if this tree trunk stood by itself and was not wedded to the vine, it would be of no interest to anybody except for its leaves. Moreover, the vine is supported by the elm to which it has been united, whereas if it had not been so married, it would lie trailing on the ground. And yet you are unmoved by the example of this tree! You shun marriage, and do not wish to wed. I only wish you would!'[14]

He tells an elaborate story – one of hundreds in the poem – of a woman's metamorphosis: Anaxarete, who was turned to stone for her hard-heartedness towards Iphis, whose unrequited love drove him to suicide. Neither this sad story, nor the analogy of elm-and-vine for marriage, has any impact on Pomona, but happily when Vertumnus stops adopting disguises, he's so good-looking she falls head over heels for him anyway. Those elms sure are handsome trees.

The mutual cultivation of elms and vines means they are often considered wedded in Italian lore. Vines were sacred to Bacchus, the Roman god of wine, fertility and fun – the Greek version of which is Dionysus – and there are many locations in Italy where grand old elms were the venue for the Bacchanalian revelries in his honour. It pleases me – as someone who spent my childhood playing under elms, indulging in illicit smokes on my elm stump at the edge of the woods – that further south in Europe in the olden days, elms were where folk went for wild, raving parties.

There are many associations between trees and alcohol. When I was poet in residence at the Royal Botanic Garden Edinburgh, we ran a boozy night sampling drinks made from the trees in the Gaelic Tree Alphabet. This ancient connection between the letters of the alphabet and native woodland species of Ireland and Scotland has enthralled me for decades. My first novel, *The Last Bear*, follows the alphabet, a chapter for each tree; I've written a poetry sequence called A-B-Tree, with a poem for each tree; and since 2011 I have been running an intermittent project using the

Gaelic Tree Alphabet as a way of engaging people in creative activities in the woods, blending scientific knowledge, folklore and literature. At least 1,600 years ago, writers of Gaelic used an inscription script called Ogham, the letters of which look like little twigs, rather like runes. The letters correspond to the first letter of tree names in old Gaelic, some of which are the same as today (for example, birch is *beithe* in both old and modern Gaelic), while others have changed beyond recognition (for example, rowan was *luis* in old Gaelic but is *caorann* in modern Gaelic). There are many mysteries about Ogham because the only remaining examples of its original use are stone and bone carvings, which are short and highly ambiguous; if it was used on more ephemeral materials such as parchment or wood, they are lost to the mists of time. I am drawn to the Gaelic Tree Alphabet simply because it's an ancient link between writing and trees, both of which are big parts of my life, and proof that the Gaelic landscape that I live in is imbued with fascinating tree culture. One of the big uncertainties and controversies is whether or not elm is in this alphabet.

The original Ogham symbols were organised into groups of five, consisting of between one and five strokes across a vertical line. First come five letters consisting of one, two, three, four or five horizontal lines to the right of the upright, representing birch (B for *beithe*), rowan (L for *luis*), alder (F for *fearn*), willow (S for *seileach*) and ash (N for *nion*) respectively. Then there are five with horizontal lines to the left of the upright, five with diagonals across the upright and finally five with horizonal lines across the upright. Gaelic doesn't use as many letters as are in the Latin alphabet, just twenty symbols are sufficient – in fact, modern Scottish Gaelic only uses eighteen letters, not requiring J, K, Q, V, W, X, Y or Z, though intriguingly it has twice as many phonemes, or sounds, as English does. All the vowels come at the end, represented by those five symbols with horizontal strokes across the vertical line: A, O, U, E, I. That final symbol with its five lines across the central stem looks just like a yew leaf, and I is indeed the first letter of the Gaelic word for yew: *iadh*. E is commonly agreed to be *eadhadh*, meaning aspen, but there's no consistency at all about A, O and U. Heather is sometimes used for U (old Gaelic *ur*) and gorse often gets used for O (old Gaelic *on*), though some versions of the alphabet differ and include broom or holly here. A is generally agreed to be linked to *ailm* in old Gaelic, but there is strangely no consensus about which tree an *ailm* might be – the two trees most commonly suggested are elm and Scots pine. Modern Gaelic is little help: pine is *giuthas* and elm is *leamhan*. The cultural connotations of these two trees are utterly different. Assuming that *ailm* translates to elm may be overly simple anglicisation; however, there are authoritative Gaelic sources who link wych elm (*Ulmus glabra*) to the letter A, including the great Gaelic herbalist Mary Beith in her book *A' Chraobh (The Tree)*.[15] Whether elm/*ailm*/A is the sixteenth letter of twenty or the fourteenth of eighteen shifted over the centuries as the alphabet evolved. Twentieth-century versions sometimes follow the conventional Latin alphabet sequence, when elm/*ailm* is found right at the start. One such is found winding up the staircase in Inverness Museum. Another is the basis of a Millennium Forest community woodland installation at Borgie Woods on the north coast of the Highlands.

The ancient Gaels frequently used trees as metaphors for people, and there are many words that double up for man or woman and tree. In poetry, heroes are compared to noble trees. For example, in an elegy for Alasdair Dubh of Glengarry, who died in 1721, the female Gaelic poet Sìleas na Ceapaich tells us that the great man was *'an darach daingeann làidir'* ('the strong, steadfast oak') and also a yew, a holly, a blackthorn and a blossoming apple.[16] The bard goes on to emphasise that the hero had no link with various inferior trees that were not considered noble: aspen, alder and elm: *'Cha robh do chàirdeas ri leamhan'* ('You had no relation to the elm').[17] When Samuel Johnson, a key figure in the English literary scene of the late eighteenth century, travelled to Scotland and claimed, despite not learning the language, that Gaelic had no literary tradition, Seumas MacIntyre responded with a poem that Gaelic scholar Michael Newton describes as 'a scathing caricature . . . which mirrors Sìleas's song but in reverse', pointing out in terms that Gaels would have easily recognised how unlike an oak or a yew Johnson was, but finishing with *'Tha do cheann gu lèir de leamhan / Gu h-àraidh do theanga 's do chàirein'* (your head is made entirely of elm, especially your tongue and your gums). Likely the wood's hardness and inflexibility are being used to emphasise Johnson's inability to form words in anything other than the English language of the oppressors.

In landscapes with Gaelic origins, there are many place names linked to *leamhan*:[18] the 'mh' is pronounced 'v', hence Achaleven (field of elms) in Argyll, Kinlochleven (head of the elm loch) in Lochaber, Loch Leven in Perth and Kinross, the River Leven in Fife, and also Loch Lomond (formerly Loch

Leamhan) in Strathclyde. Alexander Carmichael, when gathering up Gaelic folklore into his mighty collection *Carmina Gaedelica*, uncovered an ancient listing of trees, and elm gets a mention here too: '*Tagh leamhan na bruthaich*' ('Choose the elm of the brae').[19] There are many other examples to show the tree was appreciated for its elegance, even if not considered to be one of the noble species. Michael Newton urges us to note that, 'The assignment of non-noble status to particular species of the natural world does not necessarily mean that there was any attempt to persecute them or eliminate them from existence.'[20]

Still, given the many examples in Gaelic Scotland of magnificent wych elms, some of the mightiest physical specimens of any species and easily as grand as our greatest oaks, it is intriguing that the Gaels considered it to be an inferior tree. Its lesser status to the noble yew is odd, as elm's folkloric links to death are similar, although it does lack the timeless evergreen shelter of that tree. Perhaps sedate behaviour mattered to the Gaels, and elms are hasty things: in spring their flowers emerge almost as early as the hazel catkins; they are already busy fruiting by the time rowans get around to thinking about flowering; they are early to bed in autumn to make up for their early rise, with their leaves starting to golden by August, and when they drop they rot away rapidly; their seeds germinate immediately in June and their seedlings are ready for planting out the following spring; they grow fast as saplings and can be mature enough to reproduce in their teens. Their good timekeeping was useful, though: the first sign of elm's lopsided leaves in spring was said to be the signal to plant grain crops. After the dark days of winter, the elm was the symbol that the life of the new year was beginning.

But all that haste contrasts with the dignified progress of an oak tree, which can spend decades as a sapling, or a holly, whose berries can lie dormant for years before germinating, or the yew, which famously stands unchanging for literally thousands of years. The noble trees garner their girth ring by slow-grown ring, while an elm tree in good ground will have all too soon produced the wood for your coffin.

Elms have a powerful role in Scandinavian tradition, and Neil Price tells the story beautifully in his history of the Vikings, *The Children of Ash and Elm*.[21] Three gods are walking on the seashore in a time when 'there are as yet no humans in this world'. They are powerful and terrible, these three brothers, Odin, Vili and Vé, but also lonely. They find two tree trunks washed up on the beach and set about moulding them into bodies.

> First, a man – the first man – and then a woman. The gods stare down at them. It is Odin who moves now, exhaling into their mouths, giving them life; they cough, start to breathe, still trapped inside the wood. It is Vé who opens their eyes and ears, sets their tongues in motion, smoothes their features; wild glances, a babble of noise. It is Vili who gifts them intelligence and movement; they shake themselves free of the stumps, flakes of bark falling. Last of all, the gods give them names, their substance transformed into sound. The man is Askr, the ash tree. The woman is Embla, the elm . . . From this couple are descended all of humankind, down through the millennia to our own time.

Dutch botanist Hans Heybroek is well known for his efforts to breed elms resistant to Dutch elm disease – having taken up the task begun by Christine Buisman in the Netherlands – and he has also collated cultural material from around Europe relating to elm.[22] He asks the key question, 'Why ash and elm?' The answer requires us to take a step further back in the mythology. Where did Odin and his brothers come from? The answer is that they were the grandsons of giants. Their grandfather Búri, was licked out of the primordial ice by the first creature of all, a cow, Authumbla, a great hornless beast shambling about in the emptiness of the beginning of the world, whose milk sustained the first giants and gods. Heybroek points out just how important cows were for early northern people: 'It can be no accident that it was this twosome, ash and elm, that produced the essential fodder for the all-important cow, and the cow was the basis for human existence and culture . . . Out of that couple, the elm, being the most nutritive, represented the woman, while the ash, that contributed also the spear, stood for the man.' For much of Northern Europe the elm tree is archetypically maternal and an originating source of life.

The significance of elm as a source of fodder shouldn't be underestimated. Although it is a practice almost completely unheard of these days, it was normal to cut limbs full of leaves to dry and feed to livestock in winter, and elms were one of the most preferred species for this 'leaf hay'. So much so that there were taboos on the use of the wood in some places. In Norway, although elm makes good skis, it was believed you were risking your soul to do so, and if you died on elm skis you could not be buried in consecrated ground because you had already gone to hell! More positively, an elm sapling was often gifted as a wedding present because of its indirect milk-giving properties, by feeding winter cattle, and even more valued was the right to cut fodder from a mature elm. There are many legends about elms helping women who had trouble breastfeeding. This link is also found in stories about elf babies needing humans to give them milk.

In the north of Europe, elms have long been associated with spiritual beings, particularly elves, who are notoriously very long-lived, or even immortal. Elves were often believed to inhabit Neolithic burial mounds and other ancient grave sites where elms were found. According to one legend from pagan Norway, when the mother of King Olaf the Holy was due to give birth to him, the labour was so difficult that there were concerns for her life. The elm-guarded grave mound of his ancestor Olaf, elf of Geirstad, was opened up in order to take from it the dead elf's magical belt, ring and sword, which were strapped to the woman's belly, and the baby was delivered safely. The magical objects conferred upon the baby the kingship of the realm and made him into an exceptional leader, eventually being canonised for his various miracles. There are tales from Germany, Denmark and Britain about elves needing human women to help them, with midwives being called upon to attend an elf birth, usually with dire warnings about refusing food or drink while there, otherwise they may be entrapped. Similarly there are stories about elf babies needing human milk, which led to fears that new mothers were at risk of being spirited away to become wet nurses to elf children. Connections between elm and milk come up time and time again. Perhaps being upset by a wicked elf was an early folk explanation for women suffering postpartum psychosis or postnatal depression. Most stories about elves, however, refer to them as 'Good People', with the dryads, or wood elves, living quietly and secretly deep in the woods, looking after the balance of nature, carrying out powerful acts of healing or rebirth.

Scandinavians are still passionate about elms today. In 1971, a huge conflict erupted in Stockholm when the city council planned to fell thirteen wych elm trees for the construction of a new subway station. More than a thousand people established an encampment, including in the trees themselves, demanding that the subway station be moved to protect the trees. The protest was partly successful, with a new site for the station entrances and at least some of the trees left standing, and the 'Battle of the Elms' led to a new approach involving the public in Swedish decision-making – though the elm's role as a revolutionary tree really developed in America, as we'll see in the next chapter.

The Biscarrosse Elm

In the first picture I saw of the Biscarrosse elm, it has a huge, gnarled trunk topped by a few spindly branches, like a strangely distorted and outsized barnacle, with its feathery legs fluttering out when the tide is high. It is clearly already on the way towards death, its vast body barely able to conjure more than a few living twigs. This ancient field elm is crammed between buildings, a red-roofed structure on one side covered with a strange artifice of blue and white sheeting on scaffolding, and a small, log-cabin-style building right in front of it. It exists on a postage stamp of soil fenced in by a square of posts holding up a single railing like a vague attempt to stop it from running away. I've rarely seen an unhappier-looking tree, and if I were it, I would certainly consider uprooting and striding off in search of a more welcoming place to grow.

Biscarrosse is a small town in France, south of Bordeaux, close to the Atlantic coast. Its coat of arms has two red stars on either side of a tall, elegant, slim-trunked tree topped by a small green cloud of foliage – it looks nothing like the tree for which the town is famed. Like Beauly, Biscarrosse used to lay claim to the oldest elm in Europe, with local tourism information claiming it to be 750 years old just before it died in 2010, though other accounts said it was planted in 1350, making it a 'mere' 660 years old. Uniquely, every year this elm produced a ring of white foliage, described as a 'corona' or in some cases even a 'wreath'. In some versions this corona is described as 'white flowers', and in others 'leaves' – but photos are clearly of white foliage. It was, along with the mediaeval castle, a significant tourist attraction. Close to the centre of the town, it even had its own hostelry, the Café de l'Orme, and numerous gruesome stories were told about it as a place where women who had sinned were brought for punishment. There are legends of promiscuous young women made to stand naked under the tree for a day. Others tell of adulterous women being tied to a barrel under it, where they would be berated by the community. All of these stories feature public shaming of women who defied cultural norms, which gained the tree its title of *'l'arbre de la justice'*, translated as 'the tree of righteousness'.

The legend of Adeline Marsan takes us back to the fourteenth century, when Aquitaine was undergoing an extraordinary period as an English colony, isolated within France, and was the subject of endless intrigue and conflict. This local woman was accused of being unfaithful to her fiancé, a shepherd named Pierre, with one of the officers in the Black Prince's army of occupation. The Black Prince, also known as Edward of Woodstock, was the eldest son of King Edward III of England. In 1356 he notoriously captured King John II of France in Poitiers, bargaining him for rule over Aquitaine as a principality from 1362. The area, with Bordeaux at its heart, was a massively rich and significant colonial asset, not least because it gave the English access to supplies of wine and a market for grain. Edward managed to keep control of it for a decade, until the head of his army, Jean III de Grailly, was captured by King Charles V. Edward fled back to England and died four years later. This episode allows us to date the miracle of the elm's white corona to Edward's rule of the area from 1362 to 1372. The English continued to claim the area, and periodically controlled a garrison in Bordeaux, until the eventual defeat of the colonial enterprise in 1453 at the Battle of Castillon, when the French army took back control of Aquitaine.

When poor Adeline was unjustly accused of cavorting with one of the Black Prince's officers, the local elders forced her to strip naked and to stand under the elm tree. She died of shame before the day was done, and where her head had been, the tree's leaves went white in the form of a bridal crown and veil, as if to demonstrate her purity. Every year, the tree produced a wreath of blanched vegetation to remember her, and the tree became revered as a symbol of innocence and true justice, and of the culture of Aquitaine shining through the despotism of the reign of the Black Prince and subsequent English colonists.

Unfortunately, the Biscarrosse elm fell victim to Dutch elm disease in 2010, but, like so many of the best of these grand old elms, that is not the end of its story.

Some cuttings of the tree were preserved and used in a project called Bio-présence, led by Olga Kisseleva, a Russian artist who works at the Sorbonne in Paris. There, for more than twenty years, she has been leading an interdisciplinary art and science programme, engaging in and creating new forms of 'bio-art'. After the Biscarrosse elm tree died, the town's municipal council invited her to create a tribute to the tree. A profile of her project said, 'They expected some sort of sculpture. What they received instead was a miracle.'[23]

Kisseleva worked with biologists at the French National Institute of Agricultural Research, and they created a new hybrid elm using genetic material from the cuttings crossed with the Siberian elm, which is much less susceptible to Dutch elm disease. The project was a success and the result was a tree which Kisseleva describes as a 'new European subspecies' of elm, sharing the genes of these two parent species – yet another hybrid to add to elm's already baffling taxonomy.

One of these seedling trees grows where the original Biscarrosse elm stood and, most extraordinarily, apparently has the same white foliage in the form of a wreath as the original tree. Kisseleva's Bio-présence artwork celebrates the elm as both botanically and culturally rooted in Biscarrosse, restoring biodiversity to the region through its ability to nurture other species and conserving local cultural identity. Transforming the death of an ancient elm tree into art both literally and conceptually celebrates the continuation of life, just as in every ecosystem one death is never an ending but always a new beginning.

Chapter Eight

Life: American Elms

Elms are found through much of the Americas, with three species in Canada (American elm, rock elm and slippery elm), three further species exclusive to the United States (winged elm, September elm and cedar elm) and a couple more that extend southwards (Mexican elm and Ismael's elm). Elms in the Americas have been isolated from Eurasia for many millions of years, and have evolved in unique ways, so there are a number of species completely distinct from the Eurasian elms.

The most southerly, Mexican elm (*Ulmus mexicana*), is found all the way from the Mexican plateau southwards, throughout Central America to Panama. It defies the stereotype of elm as a temperate species, growing to prodigious sizes – as high as 84 metres tall – and flourishing in the tropical rainforest and cloud forest habitats of the region. It is not a favoured tree for the timber industry as the wood is hard to work, and this means that it survives unthreatened in many forests. In some areas it is prolific – in Costa Rica so much so that it lends its local name, Tirrá, to the Tirrases region around San José. Each year Costa

Rica selects a national tree of the year, as many countries do, and in 2014 a Tirrá tree from Cervantes in Cartago province was selected. Local legend has it that the tree makes the sound of a lion, roaring in the wind. Coincidentally in 2013 Costa Rica announced that the country was closing all its zoos, and the poster boy was a lion called Kivu who had been kept since 1998 in a small concrete cage in the Simón Bolívar Zoo. Environment Minister René Castro said, 'We don't want animals in captivity or enclosed in any way unless it is to rescue or save them.' Perhaps the Tirrá tree was growling in agreement.

Another predominantly Mexican elm is *Ulmus ismaelis*, named after the botanist Ismael Calzada who found it in 1997 growing close to the Mixteco River in an unusually arid limestone habitat in north-eastern Oaxaca. It also grows in Honduras and El Salvador.

The USA has three species of elms pretty much all to itself, as well as the three found in Canada, which are also widespread south of the border. Its three endemic elms are found in the warm south-eastern part of the country, with Arkansas a hotspot for elm diversity.

The winged elm (*Ulmus alata*) is also known rather delightfully as 'wahoo'. The 'winged' refers to corky growths on small branches. Unlike most elms it hates shade so it is most often found – to the irritation of farmers – growing out in the open in fields. It doesn't get huge – the national champion in Hopewell, Virginia, is less than 30 metres tall – and it has little commercial value, though its tough wood is used for making hockey sticks.

The September elm (*Ulmus serotina*) is named for its peculiar habit of flowering in the autumn. It is almost completely restricted in range to Tennessee, although the biggest one in the country is found in a cemetery in Ohio.

The cedar elm (*Ulmus crassifolia*) is also called the Texas elm – at least in Texas – though it is also found in Florida, Tennessee, Mississippi, Oklahoma, Arkansas and Louisiana. The name 'cedar' has me beat – it looks absolutely nothing like a cedar and the only explanation I can find is that it often grows in the same kinds of places as juniper. Apparently juniper is sometimes known as cedar. But just what that has to do with elm is anyone's guess. Presumably the Texans want a less peculiar name for their most common elm tree. The scientific name *crassifolia* means, roughly, 'thick-leaved' and this is at least sensible as its leaves are tougher and smaller than most elm leaves.

As we've seen, elm wood is distinctively hard, but the hardest of them all is *Ulmus thomasii*, appropriately known as rock elm, which mostly grows in the centre of the continent, from Tennessee in the south, as far west as Kansas and as far north-east as Quebec, where, just to confuse us, it is locally called *orme liège* or cork elm. This doesn't seem to be due to its use for wine corks, though the bark forms thick, corky, winglike ridges on some very old branches. These trees are slow-growing and can live – if they escape Dutch elm disease, to which they are very susceptible – for many hundreds of years, adding a mere 2 millimetres of diameter in a good year. A chunk of rock elm that was being used in a boatyard for gunwales – the curve along the top of a boat – had an incredible 250 growth rings in just 24 centimetres, so it had averaged less than 1 millimetre per year for two and a half centuries. The biggest specimen in the USA is found in Missouri, though at 24 metres tall it is diminutive compared to some of the other species. It puts up with shade to a remarkable extent, and the US government's official report on the tree says, 'Few species have rock elm's capacity for recovering from prolonged suppression,' with field evidence showing that trees have survived being overshadowed for fifty years or more without suffering, though once given light they flourish.[1] Their slow growth results in very densely packed rings and wood that is as hard as its name suggests. It grows free of knots and is both highly shock-resistant and excellent for bending when green, so it is a preferred species for many applications such as furniture, boat-building and ice hockey sticks, which need timbers to retain strength when bent into curves. It's also the wood of choice for objects that need to survive repeated impacts. It has the ability to withstand very low temperatures in life, and indeed its preference is for the continental climate of freezing winters and hot summers, which makes it almost impossible to grow in more temperate, coastal climates. Its most important attribute is the sheer beauty of its dense grain and its ability to polish to a high sheen, making it a treasured choice for lovely veneers. There's a price to be paid for this excellence, however, and the same US government report puts it bluntly: 'For this reason rock elm has been drastically overcut in many localities.' It has been given protected status in Quebec to try to give it a lasting future.

As mentioned in chapter four, one of the native elms of North America, the slippery elm (*Ulmus rubra*), has been used medicinally for a long time.[2] It is native to a large area, from Maine to Ontario in the north and from Texas to Florida in the south. It's not as large as the American elm, rarely exceeding 20 metres in height or a metre in diameter, with an open habit of growth and flattish crown. It's also known as red elm – particularly when it's cut for timber – or Indian elm, because of its traditional importance to Indigenous Americans. Other names are soft elm and moose elm, but herbalists seem to have reached a consensus on the name slippery elm, presumably due to the texture of the most-used part, which is the mucilaginous material found on the inside of the bark, hailed in the 'wellness world' as a super-herb, with anti-inflammatory, anti-oxidant and immunity-boosting properties.[3]

There is a debate around how to achieve a sustainable harvest of the bark to meet the growing demands from the 'wellness' industry. Ethnobotanists say that Native Americans sought permission from the tree, with the modest size of bark cut to make medicine ensuring scars would heal over rapidly and not be detrimental.[4] By contrast, bark-stripping on an industrial scale leaves the tree susceptible to Dutch elm disease as bark wounds are particularly attractive sites for the elm bark beetles that spread the fungus. Slippery elm is grown commercially, and plants as young as ten years old can have their bark stripped for these health products.

Many advocate slippery elm's use for good oral hygiene or as a cure for a dry mouth, but there's one use that is an essential piece of pure Americana: a spitball. One nineteenth-century technique used in baseball was to add saliva or other slippery substances, including Vaseline, to the ball to make it harder to hit cleanly. This became highly controversial and in 1920 Major League Baseball banned spitballs except for a designated list of superstar pitchers, who were permitted to keep using them until they retired. The last of these was Burleigh Grimes, who retired in 1934, but for decades afterwards many pitchers admitted to using saliva on the ball. Chewing slippery elm bark was a well-known way of generating saliva that was especially effective in lubricating the ball. Gaylord Perry, one of America's most successful ever pitchers, wrote an autobiography provocatively titled *Me and the Spitter*, in which he details his various methods for sneaking a bit of saliva or other lubricant onto the ball. Slippery elm was one of his preferred aids.

The final elm is the most widespread of all: the American elm (*Ulmus americana*). It is a deeply respected tree and a symbol of strength and resilience.[5] American elms, with their beautiful forms, were revered by Native Americans. They are sometimes called 'council trees' because councils of tribal people gathered in their shelter to make decisions. There was a long tradition of 'marker trees' or 'trail trees', most of which are now very old indeed. They act as living legacies of knowledge and cultural practices that have often been lost as the people who held them have been moved away or persecuted.

Unlike European elm, American elm keeps its leaves late into the autumn season, and this, together with its impressive size and broad canopy, suggests endurance; it's a tree you can rely on, or at least until Dutch elm disease arrives. Deeply rooted and able to withstand harsh winter weather, and indeed provide protection from it, American elm is a popular choice for public parks, churchyards and urban meeting places. North Dakota and Massachusetts have American elm as their state tree and feature many elms in municipal sites. The National Mall in Washington,

DC, is flanked by American elm groves, with continual vigilance, including sanitation felling, pruning and fungicides, to keep them healthy.

One of the most celebrated mass plantings of elms is at Colorado State University, where the famous central avenue and round park called the Oval feature about a hundred great American elms, mostly dating back to the 1880s.[6] The double rows of elms lining each side of the walkway form iconic tunnels as a result of the lovely vase shape of the trees. Their maintenance programme of pruning and disease control costs $125,000 per year, all raised from university donors.

The biggest American elm tree at present seems to be a 40-metre tall specimen on a private property in Grove City, Pennsylvania, with a girth of 8.5 metres.[7] Not quite so vast is a 35-metre tall, 6.5-metre girth elm in Deep Creek, Virginia. Canada's biggest American elm is found on Humewood Drive in Toronto, but at 'only' 25 metres tall with a 5-metre circumference, it has a way to go to become a champion.

Slippery elm, rock elm and American elm – usually known as white elm in Canada – are all native to Canada though found only in the south of the country, and again Dutch elm disease has taken a massive toll on all three species.

As in the USA, many urban areas in southern Canada have planted elm trees for their graceful shade and shelter, and of course as native species they are found throughout Canada's natural forests, except in the far north and west. Being more northerly, and therefore cooler, Canada's trees are protected from the disease in some places. However, it has ravaged populations in some south-easterly areas, such as Nova Scotia, Quebec and Ontario. In Montreal, for example, elms have been almost completely wiped out, with the death of an estimated 35,000 trees.

In Truro, Nova Scotia, when elm trees more than a century and a half old started dying, the stumps were used to make statues of famous people from the town – including early settlers, the town's first mayor, a notorious lumberjack and a police chief. These were temporary structures as many of the elm trees had rotten cores and the wood was exposed to the ravages of weather. Nonetheless, many of the statues lasted for more than a decade, some almost twenty years, and became a tourist attraction. Elms once again took a role in human history and served as an aid to our memories. When the last of them were removed in 2018, they were mourned and much missed.

Elsewhere in Nova Scotia, elms are thriving. Peter Duinker, in a review of Halifax's trees, says, 'How lucky we are that our substantial elm population in Halifax has largely escaped the ravages of Dutch elm disease. No one really knows why. Towns such as Truro and Shubenacadie were hit hard by the disease, but we have essentially escaped.'[8]

On Prince Edward Island there is a magnificent American elm, which is known simply as The Big Elm and may be the oldest specimen in Canada.[9] It is found in Glencorradale and is estimated to be 500 years old. It survived a massive fire that razed forest over a huge area of Kings County in 1727, and it was already a big tree back then. Nearly 300 years later it is still going strong. Long may it do so.

In Manitoba, millions of dollars are spent every year trying to control the disease, including tree surgery, insecticides, regular surveys and efforts to broaden public recognition and reporting of symptoms.[10] The provincial capital, Winnipeg, was home to nearly 300,000 elms when Dutch elm disease arrived in 1975. After several decades of keeping the disease under control, in recent years it has seen 'skyrocketing rates' with more than 30,000 elms lost between 2016 and 2021.[11] In 2022, 8,000 trees had to be felled. In 2023, the situation was slightly better, with just 6,000 identified as infected. Even so, it's a huge problem for the city and, as elsewhere, not cheap to deal with – the city had to allocate nearly C$6 million to the task in 2023.

Further west the disease has yet to get a foothold. In 2024, British Columbia, the most westerly province, was still entirely free of Dutch elm disease, with no cases of the fungus yet encountered, despite the presence of the European elm bark beetle (*Scolytus multistriatus*) and its vaguely stripey cousin, the banded elm bark beetle (*Scolytus schevyrewi*), both of which carry the disease in other places.[12] Elms are not native to the province, though several species are commonly planted as ornamentals. Tree nurseries in the province do a brisk trade in elm seedlings and saplings, exporting tens of thousands of them with guaranteed clean bills of health. For now, elms can breathe freely in south-west Canada and a law prevents the movement of anything elm-related into the province to help ensure this situation continues.

The most vulnerable province where the white (or American) elm is native is Alberta, with long borders to the US state of Montana to the south and to the Canadian province of Saskatchewan to the east, in both of which the deadly fungus is present. Alberta is estimated to have as many as 600,000 elms but by 2023 had only had two known cases of Dutch elm disease, in Wainwright and Lethbridge, and these were believed to have been caused by importation of firewood rather than because local elm bark beetles were carrying the fungus. In both cases, signs of the disease were spotted early and the trees were felled and destroyed before it could spread further. An outbreak in Edmonton in summer 2024 looked ominous and at the time of writing emergency felling was underway. In order to prevent the incursion of fungus-carrying beetles, there are strict rules: in Alberta (as in Saskatchewan and Manitoba), it is prohibited to prune elms, including removal of any dead or dying branches, during times of peak beetle activity to avoid open wounds on the trees, the scent of which attracts the beetles. In addition there are strict bans on transportation of firewood into Alberta, and any imported wood is confiscated.

An Alberta organisation with the marvellous acronym STOPDED, which stands for the Society to Prevent Dutch Elm Disease, offers guidance to locals. In 2023, Janet Feddes-Calpas, executive director of STOPDED, said, 'Alberta has been fortunate to remain DED free but is constantly aware of the threat of the disease pressing the Saskatchewan and Montana borders.'[13] STOPDED holds a Dutch Elm Disease Awareness Week in June each year, at the height of the disease-risk season, encouraging all Albertans to remain vigilant. They also monitor elms throughout the province using beetle traps. Feddes-Calpas said, 'Only the smaller European and the banded beetles have been found on traps throughout the province in low numbers since

1996. In recent years, higher numbers of the banded elm bark beetle have been found in the City of Medicine Hat and now are being found in more municipalities in southern Alberta.' The society has also carried out an impressively detailed inventory of elms, which make up as much as 50 per cent of all trees in many of Alberta's municipal landscapes. In 2017, this inventory estimated the value of just short of 300,000 recorded trees as C$978,149,448 – with a billion Canadian dollars resting on successfully excluding the disease, it's no wonder the issue is taken seriously. With a warming climate, the beetles' habitat is spreading steadily northwards, but it's encouraging to learn that, over a large area, disease prevention can be carried out effectively.

There is still hope that elm populations can live their lives unaffected by disease in large parts of North America. I am optimistic not only for the elms but also for us, the people who care about trees and want to live in a society that values them as fellow inhabitants.

We think of American or white elms as street trees but, of course, they also grow wild in North America, as a native forest tree, cross-pollinating over the millennia and thus genetically diverse and variable. The woods are home to the oldest and biggest of the species and there are legends of foresters sawing down vast trees for their highly valued timber but who knows how huge they grew in the past? The biggest on record in Canada was an almighty American elm tree that grew in Bruce County, Ontario, in riverside woodland near Lake Huron, on the bank of the Sauble River, which gave it its name. The Sauble Elm was 41 metres tall with a canopy spread of 26 metres and more than 7.5 metres around its trunk.[14] It was dubbed 'The Lord of the Elms' and survived for so long because when the forest was logged over by pioneers in the late nineteenth and early twentieth centuries, it was simply too big for the technology of the time; meanwhile, all around it, elms were felled, many of which were probably its offspring.

It was a tree that engendered wonder. The formal measurement of its girth was made at 6 feet up, but below that it was much broader. Malcolm Kirk, a local conservationist, described it as having 'buttresses extended out like the groynes of a cathedral'. The long life of 'the old monarch' seemed to help people imagine the vast wilderness forest that had covered the land until the arrival of Europeans. In the 1950s the first signs of disease were identified and the affected branches were pruned. This operation created a track which led to an ever-increasing stream of visitors, from botanists to school trips and holidaymakers who were wowed by its scale and sat picnicking under its shade, watching for otters in the river flowing by.

But time flowed too, and it wasn't on the elm tree's side. By the mid-1960s, Dutch elm disease was decimating the elm population in the area and the Sauble Elm was just as vulnerable as any other. Wilting in 1966 was followed by the death of the tree in 1967, and it was taken down on 2 September 1968 by a native forester, Howard McNabb, the only man in the area with the skill to handle such a vast specimen. With a couple of assistants and an onlooking crowd, the felling was ceremonial. The local paper reported that 'a tremendous roar could be heard for some distance as the elm crashed to the ground'.[15] Counting the rings inside revealed it to have been a seedling in 1701. Slabs of the wood were carefully preserved and distributed to museums and scientific laboratories but the tree is best remembered by the local people who managed to get a slice of that mighty trunk – those elm wood tables are still being eaten around today.

Since even before the Sauble Elm was a seedling, elms have been making history in the USA, gaining huge patriotic significance. As we saw with the Washington Elm in chapter one, there is often myth-making at work in these cases. One such is the Treaty Elm, long since gone, in Shackamaxon, Pennsylvania, where William Penn, founding colonist of the state, was reputed to have made a treaty in 1682 or 1683 to live in peace with the Indigenous Lenni Lenape people, who inhabited the land. The deal promised that Europeans would abide with the native people in 'openness and love' and was immortalised in a painting by artist Benjamin West, called *Penn's Treaty with the Indians*. Under a large stylised elm tree stand a bunch of frock-coated, tricorn-hatted British men who have clearly been hard at work building several impressive houses, some of which are still scaffolded, one with an unfinished roof. Two men, one on bended knee, proffer a large bolt of cloth and a box of other goodies to a group of magnificently attired Lenni Lenape people, who look greatly interested in the display. There was no written documentation of the treaty – it was a gentleman's agreement, but one that achieved great symbolic importance as a declaration of peaceful co-existence. I like to think the elm tree would have approved. In what was to become an all too familiar pattern for the Indigenous people William Penn's sons reneged on the agreement, cheating the Lenni Lenape people out of 1,200 square miles of land, so the peace it promised lasted only one generation.[16] The native people's systematic displacement left them scattered across Ontario, Wisconsin, Delaware and Oklahoma. In 1810, the elm was blown down in a storm, which feels like a fitting end to a sorry association.

In the eighteenth century, Americans gathered under an elm tree to plot a new world order. It all began in Boston, where, in the years leading up to the American Revolution, an elm beside Boston Common became the site of regular shows of defiance by the colonists against British rule. It stood at the corner of Essex Street and what became Washington Street. In 1765, protesters suspended an effigy of Andrew Oliver, a government official, from the tree. Their protest was against the Stamp Act, a tax imposed on printed documentation – a bureaucratic and financial impediment to all kinds of trade and communication. In subsequent years, the elm became the meeting place for opponents to British rule, and the space under its canopy became known as Liberty Hall. Regular gatherings were held there, and the ringleader of the resistance, Ebenezer Mackintosh, was referred to as Captain General of the Liberty Tree. When the Stamp Act was repealed in the following year, the rebels celebrated by decking the elm with ribbons and flags. Soon the tree was used for unofficial trials, including that of a customs boat, which was found guilty and burned in 1768 as part of the Liberty Riot, and that of a customs official, John

Malcolm, who was tarred and feathered under the tree in 1774. The British forces, fully aware of the symbolic importance of the elm, considered it to be such a threat that they felled it in 1775 and chopped it up as firewood.

Thomas Paine helped to solidify the reputation and symbolic value of the liberty elm. A British-born revolutionary who was instrumental in both the American and French Revolutions, and one of the Romantic Era's most important political writers, he had moved to the American colonies in 1774, where he published a pamphlet, *Common Sense*, calling for independence from Britain and arguing for a democratic and egalitarian basis of society. He also wrote a ballad, 'Liberty Tree', in the year the Boston elm was cut down by the British troops.

> In a chariot of light from the regions of day,
> The Goddess of Liberty came;
> Ten thousand celestials directed her way,
> And hither conducted the dame.
>
> A fair budding branch from the gardens above,
> Where millions with millions agree,
> She brought in her hand as a pledge of her love,
> And the plant she named Liberty Tree.
>
> The celestial exotic struck deep in the ground,
> Like a native it flourished and bore;
> The fame of its fruit drew the nations around,
> To seek out this peaceable shore.
>
> Unmindful of names or distinctions they came,
> For freemen like brothers agree;

With one spirit endued, they one friendship pursued,
And their temple was Liberty Tree.

Beneath this fair tree, like the patriarchs of old,
Their bread in contentment they ate;
Unvexed with the troubles of silver and gold,
The cares of the grand and the great.

With timber and tar they Old England supplied,
And supported her power on the sea;
Her battles they fought, without getting a groat,
For the honor of Liberty Tree.

But hear, O ye swains ('tis a tale most profane),
How all the tyrannical powers,
Kings, Commons, and Lords, are uniting amain,
To cut down this guardian of ours.

From the East to the West blow the trumpet to arms,
Thro' the land let the sound of it flee;
Let the far and the near all unite with a cheer,
In defense of our Liberty Tree.

There is nothing specifically elm-like in Paine's poetic tree, perhaps deliberately focusing all its rhetorical energy on the symbolic nature of the plant as a representation of freedom from tyranny. Yet it turns the felling of a tree into the cutting down of a guardian spirit, the response to which is a call to arms. This transforms a mocking move by British soldiers into a military provocation that challenged the American people to fight back. The elm as a symbol of liberty was instrumental in the start of a war.

Like so many significant elms, Boston's Liberty Tree has lived on in the cultural memory of the nation, first with a post, then a plaque, then a sculpture, then a small park and these days a full-blown Liberty Tree Plaza with a monument, seats and, of course, a replacement elm tree. Liberty elm is also the name given to a hybrid elm variety, partly resistant to Dutch elm disease, that was distributed by the Elm Research Institute from the 1990s.

The cult of liberty elms was by no means confined to America. Revolutionary groups gathered under elms throughout Europe, and they were planted as symbols of liberty in France, Ireland, the Netherlands, Belgium and Italy. In the heady days of the French Revolution, a new calendar was introduced, and the twelfth day of each month was declared *'le jour de l'orme'* ('the day of the elm'). In 1790, inspired by Boston's Liberty Tree, L'Orme de La Madeleine (the Elm of La Madeleine) was planted in Faycelles, Département de Lot, as a liberty tree, and it stands to this day. In Calabria in 1799, L'Olmo di Montepaone, L'Albero della Libertà (the Elm of Montepaone, Liberty Tree) was planted to celebrate the new republic. And in 1839, after the Greek Revolution, a thousand elms were planted in the national garden in Athens.

One of the largest ever American elms grew in the state of Ohio, near Circleville in Pickaway County. Under it, in 1774, a Native American chief called Logan made a heart-rending proclamation, as follows:

I appeal to any white man to say if ever he entered Logan's cabin hungry, and he gave him not meat; if ever he came cold and naked, and he clothed him not. During the course of the last long and bloody war, Logan remained idle in his cabin, an advocate for peace. Such was my love for the whites, that my countrymen pointed as they passed, and said, 'Logan is the friend of the white men.' I have even thought to live with you but for the injuries of one man. Colonel Cresap, the last spring, in cold blood, and unprovoked, murdered all the relations of Logan, not sparing even my women and children. There runs not a drop of my blood in the veins of any living creature. This has called on me for revenge. I have sought it: I have killed many: I have fully glutted my vengeance. For my country, I rejoice at the beams of peace. But do not harbour a thought that mine is the joy of fear. Logan never felt fear. He will not turn on his heel to save his life. Who is there to mourn for Logan? Not one.[17]

Logan was a chief of the Mingo, an amalgamation of depleted tribes who joined together to survive. His father was a powerful chief and diplomat, Shikellamy, who formed an alliance with a Pennsylvanian official, James Logan, from whom his son took his name. At first, Logan followed his father's example in seeking peaceful relations with white settlers, but his people were lured to a cabin at the mouth of Yellow Creek, where all of them, except for one mixed-race child, were killed. Although the perpetrator was thought by Logan to be Colonel Cresap, his descendants attribute it instead to Daniel Greathouse, leader of a band of aggressive 'long knives', a tribal term for belligerent Virginian settlers. Whoever was responsible, after the slaughter of his family, Logan was driven mad with grief and sought bloody vengeance. Retaliation for murder was part of Native American custom, so other Mingo warriors joined his rampage. When Dunmore finally sued for peace, after military reprisals, Logan had little choice but to welcome it.

Some versions of the story claim the treaty was negotiated under this huge elm's shade, though others say that it is only where Logan approved its terms. Modern pictures show a mighty plant with a trunk more than 7 metres in girth and a canopy stretching 50 metres. Although it sickened with Dutch elm disease and succumbed to a storm in 1964, it played such an important role in American history that it will live on in story, and it has been replaced by a new, younger elm to stand as a reminder. The wisdom that emerged under that Ohio elm went right to the heart of the problem of colonisation.

A century and a half after Logan, in 1941, a local man described as the Ohio poet laureate, Frank Grubbs, wrote the poem 'The Logan Elm', which includes the lines:

> O, ancient elm! When Logan stood
> Beneath thy kindly shade
> He little dreamed his eloquence
> For him a shrine had made

I imagine those 'beams of peace', which Logan rejoiced in, as sunshine filtering down through the elm tree's foliage. America's Indigenous people tried to make space for white settlers but were so brutally abused that they were driven to violent self-defence, yet despite persecution, disease, loss of land, compromised identity and terrible personal grief, they had the magnanimity to welcome peaceful solutions. Logan's speech is rightly remembered for its humanity, for the acknowledgement it makes of horrific wrongs perpetrated on both sides, and for the way it shows that even the most personal of conflicts can be transcended. In the tree's long life, it saw many horrors but stood through it all as a shrine for peace.

At 33 metres tall and 6 metres round, Herbie was by no means a record-breaking tree, though it was known as 'the tallest elm in New England'.[18] Yet this elm, planted in the 1770s in Yarmouth, Maine, and eventually felled by a storm in 2010 after fighting off multiple bouts of Dutch elm disease, lives on through the Liberty Tree Society. This national society of elm fans has created clones of the mighty tree, which was believed to have unusual resistance to the killer fungus, and these are sold far and wide in an effort to replace some of the millions of lost elms. Herbie's story would be incomplete without Frank Knight, Yarmouth's tree warden, who spent a large part of his career protecting the tree, who he described as 'an old friend'.

Once Dutch elm disease arrived in Yarmouth, Knight realised that most of the elms under his care were doomed and so he focused his efforts and attention on the tallest, oldest and most handsome tree. Through successive waves of infection, he pruned it, applied pesticides and fungicides, and managed to keep it alive, all while looking for ways to sustain its genes. He was aged 101 himself when the veteran elm finally had to be felled, standing by as if at the bedside for the passing of a loved one. Before the demise of the tree, he worked with John Hansel from the Elm Research Institute to take many thousands of cuttings to propagate them, so Herbie lives on in many places. The love of this one man is reflected in widespread affection and respect for the tree. Just as many people leave their body to medical research after their death, some of Herbie's wood was 'gifted to science' after its demise.[19] Specifically it was passed to dendrochronologists, led by Peter Lammert of the Maine Forest Service, to unravel the story hidden in the tree rings. The tree, unusually solid to its 212-year-old core, has revealed past years' weather patterns, increasing our understanding of the changing climate.

The Survivor Tree grows at the site of a bomb blast in Oklahoma City at the Alfred P. Murrah Federal Building. On the morning of 19 April 1995, Timothy McVeigh, ex-soldier and American domestic terrorist, parked a rental truck full of explosives in front of the building. The explosion killed 168 people and injured hundreds more; it was the worst act of terrorism the USA had ever seen until the attacks on the Twin Towers in New York on 11 September 2001, and it remains the deadliest domestic terrorist attack in the USA ever. McVeigh and several accomplices were captured and prosecuted, and McVeigh was later executed. Extraordinarily, although the Murrah building was virtually destroyed and needed to be dismantled, and hundreds of nearby buildings also sustained damage, the elm tree growing right outside, estimated to be about a hundred years old, survived the blast. It's a wonky specimen, about 15 metres tall, with its main stem not quite vertical, a fork in its trunk about three metres up with one side accentuating the lean of the tree, and a big, almost horizontal branch off the main stem that counterbalances it. The overall effect is far from elegant, but its lopsided, quirky character is fitting for a tree that stands in memoriam, as if knocked sideways but still standing, offering shade and shelter to those who come to mourn those 168 victims, each represented by an empty chair in the courtyard where the horror struck.

The Survivor Tree nearly didn't survive at all, though the greatest threat wasn't the bomb. Immediately after the blast, the FBI wanted to cut it down to retrieve shrapnel or other evidence that might be wedged in its wood. Fortunately, however, its significance was soon recognised when President Clinton stood beside it to make a speech, knowing the symbolic power of elms. The elm became a kind of shrine, with flowers and wreaths laid beside its trunk, and the FBI had to get what evidence it could without harming it. The Survivor Tree was a central part of the design of the memorial, with a commitment to its protection written into the mission statement: 'As preliminary planning for a permanent Memorial began, it was quickly determined that any design must include the Survivor Tree – an integral part of the story of what happened here, as well as our hope for the future.'[20] As a 'beacon of hope', seeds and cuttings of the Survivor Tree are harvested each year and offered to the many wounded survivors of the blast, family members of the victims, emergency service staff and local schools. Elm is playing its traditional role as the tree of life after death, giving comfort to those who are bereaved, offering hope for a better future to those who find themselves broken by loss.

My visits to grand old elms sometimes feel like making friends at a nursing home. I've known, since the disease arrived, that the chances of many of these trees surviving more than a year or so is slim. Assuming the usual rate of death applies here, leaving 1 per cent of the trees, our parish will have only a handful still standing after Dutch elm disease has run its course. It is difficult to remain hopeful when more often I feel panic and outright distress at the idea of our lovely big trees being doomed. Yet there are positive actions we can take: watching out for early signs of disease and pruning diseased limbs before the whole tree is affected; burning any elm logs in our wood stores that might be harbouring eggs, grubs or fungus; gathering infected wood into a central location where it can be debarked, chipped, heat-treated or burned; cleaning any tools used to handle elms to avoid spreading the spores; above all, encouraging people to care. Most importantly, we mustn't give up on elm. We have to plant young elms, to ensure that the next generation can carry on, so as I was drafting this book, I began planning a tree-planting event with the local primary school children, the ranger service and an artist who would get everyone painting seedling portraits. Most excitingly, the event would also include a ceremonial planting of a special elm seedling that had been bred by the Royal Botanic Garden Edinburgh from elm trees further south in Scotland that had lived through successive waves of Dutch elm disease and seemed, somehow, to be able to remain healthy.

This seedling was part of a programme to propagate 'potentially resilient' elms and distribute them to good elm country around the Scottish landscape in the hope that their ability to avoid disease will spread through intermarriage with local trees, increasing the resilience of our overall elm population.

The manager of our local community tree nursery, Nick Clooney, and his assistant Josie, were keen to help. They had a polytunnel half-full of young elms, all grown from the seeds of our local trees, which they'd been collecting for many years. Nick had sold young elm trees to local tree planting projects every year for the past decade. In one such scheme, seven-year-old elms were thriving in an ideal habitat alongside a stream bank. This wasn't the beginning of our rescue mission: it was already well under way.

I asked Nick and Josie to keep three elm seedlings aside for my own deer-fenced exclosure on the croft. When I went to see them, even though out in the wind all the trees were bare of leaves, inside the shelter of the tunnel the seedlings still carried their little saw-toothed edged flags of green, defying the season, so enthusiastic in their mission to grow. Standing over them, bending down to them, it was as if the oxygen they exhaled was a special form, steeped with optimism. I breathed in lungful after lungful. Small they might have been, but there were hundreds of them, and each one a potential tree. There, in that one bed, as many trees were rooting as the adults I was so worried about out in the crags and ravines of our parish. In those few square metres, I found what hope felt like. I breathed it in.

A Hebridean Elm

The Assynt elms are the most north-westerly on the UK mainland, but there are elms further north-west on the Hebridean Isle of Lewis, next stop America, and one of them is my own Survivor Tree. On a good day, we can see Lewis from our croft, and I sail across the Minch to the Outer Hebrides at least once a year. It's a wild stretch of sea, open to the north and susceptible to heavy weather from the Arctic. The ocean remembers weather and transmits its effects through gentle or violent undulations. Unfortunately this swell from the north is often at odds with the local prevailing wind, making for a 'confused sea' and horrible sailing, particularly if the tides are strong, tossing the boat about sickeningly as each wave hits from a different angle, sometimes jostling her from both sides at once. We must pick our days for a Minch crossing carefully.

In the winter of 2023, we have ferocious storms, and more than one from the east. Easterly winds tear across Scotland, funnelling between the mountains to emerge onto the west coast in fits and starts, lulling and then gusting unpredictably and often with immense force. Storm Gerrit in December 2023 brings local winds of 89 miles per hour, hitting our little yacht, along with five others, on the pontoon in Lochinver harbour, causing damage that we are unable to fix locally. In January, when the sea is again benign, we wrap up, set off early to catch all the sunshine and sail across the Minch to arrive in Stornoway – Lewis's main town and the capital of the Western Isles – just as dusk is falling.

Stornoway is as friendly as a village and boasts Lews Castle, with magnificent wooded grounds, right beside the harbour, and in those grounds there are lots of elm trees! Next morning, I wander the January woods and initially take lime trees and sycamores for elms as they sometimes produce bushy growth around the base of their trunks in a similar way to an elm but up close, sycamore twigs are far too chunky for wych elm's delicate ones, and the bristles around the boles especially so. Lime's buds are russet-red and, although small, not tiny enough for wych elm, whose buds are glossy brown, like a horse chestnut conker only so much smaller.

On a flat piece of ground a host of elms tower up among beeches and sycamores, all striving for the light, producing a mutually supporting canopy. In the morning light on this cold winter's day, sunrise turns the entire canopy bronze, as bright as if the trees were still carrying their autumn leaves. Massed trees are magnificent, and any one of these elms is probably taller than the biggest of all those in Assynt, out of necessity to compete with its fellow woodland dwellers. However, this striving for light means that all these trees are straightish-trunked, high-branched and relatively nondescript. I guess the soil is fertile and their life of relative ease has made them perfect forester's specimens. I've grown used to knobbles and gnarls, oddity brought about by tenacious gripping in unlikely places, and none of them has the personality of my local trees. On our croft many of them are wind-blasted into unlikely forms and I have a deep fondness for the kinds of trees that timber-seekers find worthless. The elms I grew up with, in my childhood Northumbrian wood, clung to the sides of a steep ravine. They, too, were all kinds of interesting shapes, with big horizontal trunks leaning out over the precipice or multiple stems from a single huge bole. Perhaps grazing animals once roamed freely there, or were let loose to eat the beech mast and acorns, and the only elms to survive their teeth were inaccessible.

Eventually I find the tree that catches my heart. Although as tall as its neighbours, it has a huge wound in its trunk, a gaping hole from its roots up to about 2.5 metres above ground. The bark has grown back and looks as if it has melted and flowed inwards to heal, and then solidified. Inside, the wood is papery with rot, latticed with fungi and spiders webs holding dust from the workings of boring insects. This rotten inner wood forms a complex pattern of black, bronze and off-white, giving a charred effect, with glowing embers and ash. Whether the damage was caused by a fire or it was simply debarked by a grazing deer while young, I'll never know, but it has survived some trauma in its youth, lives on and flourishes. The remaining two-thirds of the circumference of the tree is mossy and unscathed, and the fissures in its bark flow like the patterns of foam on the river below, where our boat is moored.

I return to Stornoway repeatedly as our boat repair progresses. In mid-March, when it is not yet the equinox, this survivor elm is flowering already. Its lowest branch is many metres above my head and the few visible twigs have only leaf buds, still resolutely closed. All the sport is up in the canopy. I crane my neck, greeting two crows on high branches. The delicate lattice of winter has turned to lacework, with each twig adorned by a little sprig of growth, a fluffy bobble of flowers, as if a calligrapher has drawn filigrees on all the letters of her script.

Around the tree, the first signs of spring green are emerging slowly: celandine leaves, grass making a first effort at growth, ground elder much more enthusiastic, a cress that will no doubt double in size daily once it starts thinking of flowers. Other than the decorative daffodils planted beside the shoreside path, in the woods only the elms are in bloom, all covered in little pompoms, up there high in the sky. It is bitterly cold, maybe 5°C, the wind picking up ahead of a rain squall. My survivor elm is a hardy – perhaps foolhardy – tree, but I love it. Unlike its straight neighbours it is curvaceous and lumpy, tall, slim, unstraight, almost wiggly in its upper twigs and branches with their exuberant fistfuls of out-of-reach flowers dancing with the spring breezes, daring the spring gales, busy already with the new season's breeding. And there are dozens of other elms here to share pollen with and to catch what they offer in return.

A big camera lens reveals the flowers close up. Each twig sprouts little brushy balls of delicate structures – dark bud scales, like beetle carapaces, burst open to allow a whole cluster of little flowers to blossom. The hard brown casing contrasts sharply with the softness of the interior: each flower has a pale green, pink-tinged sheath, as light as a petal, although botanically this is the sepal and there are no petals. Sprouting out of this sheath are four or five stamens, pale fleshy stems each bearing at their tips a pollen-rich anther, a little blob like a match head that matches the dark colour of those bud cases. This is the male sexual organ of the flower, archetypically erect and outreaching. Beneath these stamens is the female part, a Y-shaped stigma, like two hairbrushes lying ready to sweep up any pollen that is blown within reach. There is a red tinge to the brown of the bud case and anthers, and with the pink sheen of the sepals the overall effect is of a blush of colour and a blurring of sharp edges as each twig sprouts its fluff in preparation for sex to begin.

At first, only the female part is ready for action and it will take a few more days for the anthers to start flinging their pollen to the wind. This is a tactic to give the female parts some time to get exposed to genetically different pollen from other trees because they can't be pollinated by their own male flower parts. They recognise and reject the pollen from their own tree, which is called 'out-crossing', to maximise genetic diversity, though it makes life difficult for secluded elms or those surrounded by clones. Here in Stornoway, I'm hoping that the many fine elms in the wood, including my survivor tree, will find themselves dusted by fine-smelling pollen from some satisfactory stranger.

By April, our boat repair is done and we haul her out for painting while the weather is fair. The elms are forming samaras. They, and we, are readying to sail, to let the wind take us. What greater symbol of survival and freedom could there possibly be?

Chapter Nine

Death: Elms in the Arts

Poems about elm trees are often also poems of grief. This is hardly surprising, given that the tree has been a symbol of death in so many cultures, and how uniquely devastated they have been by disease. There is growing recognition of how pervasive grief is, including the deep societal heartbreak about ecological losses and damage to places we love. Sociologists and psychologists have started studying the detrimental impact on our mental health of 'environmental grief', which social work researcher Sandra Engstrom neatly characterises as 'recognition that ecological destruction also means our own destruction'.[1] We are part of nature, and just as we can be shattered by the death of a member of our own family, when the natural world is ruined, we can find a part of ourselves uprooted. Engstrom recommends addressing the loss by 'seven generation thinking', a technique she learned from North American First Nations. Over seven generations of people, a burned forest can regrow, a quarry can become a lake, a volcano can become swathed in lush green vegetation.

Hope for elms can be found by a shift in temporal scale. We rarely think about what will happen beyond this year or maybe next; we vote politicians in for five-year terms and are constantly occupied in making decisions about the immediate short term. A bereavement may shock us into reflection on the whole course of a human lifespan, but even so, we are inclined to revert to platitudes about 'living every day to the full' and seek cures to grief that involve staying resolutely attentive to the present moment. But even a whole human lifespan is a small fraction of that of an elm tree: the Guardian of the Gateway in Beauly lived almost ten times as long as either of my parents. Moreover, when we start to think about how disease-resilient genes might spread among an elm tree population, we need a perspective that sees ahead not merely hundreds but many thousands of Earth's circuits around the sun. Art and literature can provide that kind of perspective.

Psychiatrist Elisabeth Kübler-Ross famously claimed that there are five emotional stages of grief: denial, anger, bargaining, depression and finally acceptance.[2] This theory is flawed and a poor fit to many people's experience (as I attempt to endure the fallout from my father's death, these five stages bear little resemblance to my life), yet it endures, perhaps because something as baffling and all-encompassing as death makes us grasp for structures that can make sense of it. I have borrowed it as a model for roughly grouping poems about elms according to the different emotional responses they evoke.

DENIAL

Whether English poetry made elms paradigms of the English landscape or whether English poets were responding to the trees' pre-eminence, there are countless references to elms in English pastoral verse, many of which are simply appreciative or otherwise implicitly deny it is in peril or linked with death. One of the most famous is Shakespeare's use of elm in his play *A Midsummer Night's Dream,* in which he evokes the classical image of a male elm entwined by a female vine or ivy, in the words of Titania, the fairy queen, in her rapture about Bottom:

> Sleep thou, and I will wind thee in my arms.
> . . . the female Ivy so
> Enrings the barky fingers of the Elm.
> O, how I love thee! how I dote on thee![3]

Robert Browning's 'Home-Thoughts, from Abroad' opens with what has become a spring cliché: 'Oh, to be in England / Now that April's there'. The first image he gives us of the spring morning he is missing is well-observed: 'the lowest boughs and the brushwood sheaf / Round the elm-tree bole are in tiny leaf'.[4]

Another often-quoted classic elm reference is found in Rudyard Kipling's 'A Tree Song', which works its way through all the most common trees of 'Old England' and some of their folklore. I'm not sure whether this is the actual source of the 'old saying' of 'Elm hateth man and waiteth' or if Kipling was quoting directly from the lore. Either way, the poem goes on to explain the branch-drop tendency that the saying is referring to:

> Ellum she hateth mankind, and waiteth
> Till every gust be laid,
> To drop a limb on the head of him
> That anyway trusts her shade.[5]

In his characteristically radical way, with over-the-top alliteration and made-up compound words, Gerard Manley Hopkins shows us the way an elm tree produces its beguiling dappled shade. In a poem with the unlikely and rather unhelpful title of 'That Nature is a Heraclitean Fire and of the comfort of the Resurrection', he writes: 'wherever an elm arches, / Shivelights and shadowtackle in long / lashes lace, lance, and pair'.[6] I love the way this presents us with the light and shade under the elm as if they are lovers, jostling, jousting and joining together.

Robert Frost was a big fan of elm trees, particularly those that he found in Britain. One of his poems, 'The Sound of Trees', was inspired by a group of elms growing near a house he lived in at The Gallows, a cottage in Dymock, Gloucestershire, between 1912 and 1914.[7] According to John Haines, a Gloucestershire friend, 'The trees seemed to be speaking to him about difficult decisions that had to be made as a result of the war.' These elms are very much rooted, described as something 'that talks of going / But never gets away' and ultimately 'means to stay'.[8]

There is also an 'aged elm' in another much-loved Robert Frost poem, 'The Need of Being Versed in Country Things'. It stands by the ruins of an American house and barn that have been destroyed 'in flame'; the elm itself was also 'touched with fire' yet, like that mighty elm that stands on Prince Edward

Island despite the forest fire that raged around it, the elm in Frost's poem remains.[9]

If you like tree poems, then Angela King and Susan Clifford's anthology *Trees Be Company* is essential reading. In it, Geoffrey Grigson's poem 'Elms Under Cloud' presents us with beautifully old fogey English elms as 'old men with thinned-out hair, / And mouths down-turned', which 'express / The oldness of the English scene'.[10] Norman Nicholson, in the same anthology, fills us in with the long history of wych elms in the northern English landscape. 'The Elm Decline' covers 10,000 years to remind us that the trees arrived on a post-glacial 'tidal surge / of oak, birch, elm' and were then, five millennia ago, subjected to Neolithic axes, which 'ripped the hide off the fells'. The implication is that the present-day death of elms is nothing out of the ordinary. These are just two of seven poems that are entirely devoted to elm trees in this short anthology from the late 1980s, and there are numerous passing mentions of elms in other poems, most of which are by British writers, so overall elms have a presence in the book akin to their place in the British landscape of the last century.

In a more recent anthology, *The Tree Line: Poems for Trees, Woods and People*, published in 2017, there isn't a single elm poem amid all those about oaks, ashes, beeches, birches, rowans, willows and alders, and in the entire book elms only get two passing mentions, one in 'Laws for Trees' by Alyson Hallett, within the morbid phrase 'Elm vs Saw', and the other in a poem by Peter Robinson, 'Die Holzwege' (the title of a book by Heidegger, usually translated as 'Off the Beaten Track'), which says: 'Dutch elm, dieback, acid rain / find the woods in trouble . . .'[11] Elm has largely disappeared from the British poetry landscape, as if we're in a collective denial of its former significance. Only its ghost is found in another new anthology, *The Book of Tree Poems*, compiled by Ana Sampson in 2023, in an old poem 'Child's Song in Spring' by Edith Nesbit, who has been dead for a century. In her day, she said, 'The elm tree makes the old churchyard shady,' so even back then it was well and truly the tree for mourning.[12]

In the *Trees Be Company* anthology, there are a couple of very short but thought-provoking elm poems that make no explicit reference to their demise. Hilaire Belloc's 'The Elm', in six lines, places a girl, Dorothea, in a landscape of a corn field and a single elm tree, the simplicity of which evokes a child's painting. Dorothea smiled, not just once but in the first, third and final lines. The poet says that neither he nor she knows why, but the title of the poem and the sparseness of anything other than that tree imply that when 'A sudden glory had bewitched the child', it is the elm that made her smile.

Even shorter and more personal is John Fuller's 'The Elms'. In the first short line of four, he places us in the cool shade of the trees, then rain comes and is 'heard in the great elms'. He makes it clear that every single drop of rain and every leaf of the tree is significant. And in the final line, an epiphany: 'I am still alive.'[13] Whether this is by contrast with dying trees, or like the living ones he is standing under, life can by no means be taken for granted.

Genevieve Taggard, with her 'Dilemma of the Elm', comes closest to expressing denial about this loss. Both the opening and penultimate lines are, 'In summer elms are made for me,' so she clearly holds them close, but 'I walk ignoring them,' she says in line two, and she closes the poem by ignoring them again. They ignore her back, and she likes the elegant way they do so. Between these two summer face-offs, the poem presents us with winter in which the elm becomes 'a wiry fountain' and she hears 'the lonely scrape / Of rooty branches', which makes her doubt her memories of their leaves. But this very disbelieving thought brings colour to her mind, and the summer elm returns superimposed upon the winter one – 'I see it green, absurdly new' – so this winter elm is 'a double tree'. This combination of the reality of the bare wood and the memory of its foliage is unsettling, in just the same way that the skeleton of a diseased tree reminds us of the loss of life, and she admits that 'all elms trouble me'. But no sooner does she concede this than she's back to summer, back to the tree being made for her, and back to ignoring it all over again. This poem, with the elm as a symbol of death, shows us how we can barely dare to look at our mortality square on.[14]

ANGER

The English poet of the countryside, John Clare, makes several references to elms and when he turned his attention fully to the tree, the result was a hugely powerful piece, not surprisingly because the beloved tree was felled as part of the process of putting what had been common land to more profitable use by the landowners. In 'To a Fallen Elm', Clare stands head and shoulders above all other poets in his rage against the felling of a tree. In mourning, Clare treats the downed elm as a symbol of the social injustice that shifted land use from stewardship by poor people to the aggrandisement of the rich who 'barked of freedom' while riding roughshod over the local community. Clare spends the first half of the poem declaring his friendship with the tree, then shows us the 'knaves' and 'hypocrites' who sheltered from weather under it then cut it down. The regularity of his metre seems to echo the rhythm of the axe that 'felled thee to the ground . . . O I hate that sound'. And with this declaration of hatred he describes how 'the common heath became the spoilers prey', with poor people ending up in the workhouse prison and nature overrun by agricultural development. He ends as follows:

> Such was thy ruin music making Elm
> The rights of freedom was to injure thine
> As thou wert served so would they overwhelm
> In freedoms name the little that is mine
> And these are knaves that brawl for better laws
> And cant of tyranny in stronger powers
> Who glut their vile unsatiated maws
> And freedoms birthright from the weak devours.[15]

We can picture Clare among the ruins of the great old tree, giving vent to what starts as grief then escalates to righteous fury.

Clare's poem inspired another wonderful piece, 'The English Elms' by Carol Ann Duffy, which does not burn with the same fiery rage but simmers angrily at the cultural and social losses from the death of the trees. Opening with the 'Seven Sisters in Tottenham, / long gone', Duffy gestures to elms lost from all our towns and cities and the rural landscape where their shapes echoed clouds, 'like green rhymes'. A whole generation are now strangers to their forms that 'loom' out from old films, 'or find posthumous fame / in the lines of poems'. It is left to writers to conjure elms back into 'a world without them', to perform that magic trick of defying time that the written word can do so well. Her poem shows all the others who would have used elms and who have also disappeared: the woodcutters who used the wood for coffins, the boys who played cricket with elm stumps, the painters and artists who made their portraits, and all the other life who benefited from them – the sheep who sheltered under their leaves, the birds who nested in their branches. On she goes until she peters out with her last reference to the trees, followed by an ellipsis – 'elms . . .' – as if she could never run out of things to remember, and has to pause, to take a breath of disbelief at the horror of what she has to finally admit they were: 'great, masterpiece trees, / who were overwhelmed'.[16]

BARGAINING

'The Tree Agreement' by Elise Paschen describes a neighbourhood spat about an elm tree that overhangs from the poet's garden.[17] Her next-door neighbour 'demands we hack / it down'. It's a Siberian elm, presumably growing in Chicago, where she is from, and the neighbour claims it to be 'a "weed" tree'. As explained in chapter six, Siberian elms in some parts of North America are spontaneously spreading and considered by some to be invasive. The neighbour says its 'leaves overwhelm' his side of the boundary. But the poet's family see a helpful tree that screens unsightly things like ugly buildings and railway lines, providing shelter and shade and a 'leaf-hoard' habitat for other species:

'crossway / for squirrels, branch house for sparrows, jays' whose 'chatter-song drowns out cars below'. They resist the neighbour's demands: 'We disagree, / claim back the sap, heartwood, wild bark.' They exhort the tree to 'root deep'. As the neighbour says, 'leaves overwhelm', but for the poet, that is a reason for the elm to remain. She likes being overwhelmed by elm.

Another American poem, 'Ever Vigilant: Julius V. Combs, MD' by Melba Joyce Boyd, begins with Dr Combs' date of birth in 1931 and death in 2020, so the poem itself is a memorial. It starts with the elm's inspiration, 'sheltering Hazlett Street', where Dr Combs was presumably born. The poem is long and thin, with short lines, as if in imitation of a tall, elegant tree. Its second stanza honours the doctor's commitment to medicine, which it credits to him 'reflecting on the / arc of tree limbs, / supplying fresh / oxygen for lungs'. When the elms died of disease 'leaving empty skies / and abandoned homes' the poem portrays a neighbourhood left utterly desolate and 'hemorrhaging like / bodily injuries'. Yet the good doctor works on, continuing to give himself to the people bereft of their trees, and thereby 'resurrecting a / legacy' that trades-off their loss with good deeds, and expresses a spirit of defiance not only of the degraded environment in which people must live, but of the racism so many Black people have had to suffer.[18]

DEPRESSION

As an angst-riddled teenager, I often stomped out of the house wearing my wellies and favourite old green jacket, and made my way over the field to the woods, biting back tears or simmering with rage or just feeling down in the dumps. Through the kissing gate at the far side of the field, with no one to kiss, I turned south and followed the badger-beaten path along the fringe of the stream gully at the end of which was my elm stump. There I sat on its smooth rings of comfort, rooted to the earth, maybe smoking a cigarette stolen from Mum's handbag or rolled from some tobacco bought with hard-earned pocket money. Often I'd just look at the River Tyne flowing by, far below, from the ruined lime kiln chimneys to the disused railway bridge at the bottom of Hagg Bank. Sometimes I'd have a book in the commodious pouch pocket of that old cotton jacket. From my early teens, poetry was a favourite. I loved e. e. cummings for his nose-thumbing at convention and I was entranced by Sylvia Plath, like many young women who find she speaks so vividly of being engulfed by emotions that are almost inexpressibly unbearable. I was already starting to experiment with poetry, and on the inside front cover of my first notebook, which I titled 'Young Witch Scribbling', my thirteen-year old self inscribed the tenth and eleventh stanzas of a Plath poem, about a malignant, feathery entity that 'flaps out' every night in search of 'something to love'.[19]

I held this quotation in my memory for decades, losing track of the poem itself and forgetting that this owl-like creature flew out into the darkness in search of prey from an elm tree. The identification I felt with the felled stump, which seemed to understand all my horrors and woes, is mirrored in Plath's 'Elm'. This powerful poem centres on an elm tree which spoke with certainty and galloping energy to all the fears of my teenage self, her sense of brokenness and most of all to her longing for love. The tree itself is female, we know from the first line, with its opening that couldn't be less ambiguous about the mood of the poem: 'I know the bottom, she says.' It is unremitting in its misery. By halfway through, even before she gets to the terrifying 'dark thing', she has suffered from the sun, from wind and from the 'merciless' moon, with whom she has tangled, leaving either her or the moon or both of them 'Diminished and flat, as after radical surgery'. By the end, the mystery is absolute and devastating: a 'murderous' face, which she is 'incapable' of knowing, 'petrifies the will'. The final line fells us with three axe blows: 'That kill, that kill' – and just to make sure we've got the point, bringing in the typical mythic power of something spoken thrice, like a magic spell – 'that kill.'

In eight long lines, C. K. Williams' 'Elms' is similarly heart-rending. It opens one morning with the 'tree men' dealing with the 'stricken' trees. In its second line we hear them, so vividly, as the tragedy unfolds: 'The pitiless electric chain saws whine tirelessly up and down their piercing, operatic scales'. After each tree falls, its debris is shredded and 'truckload after truckload' is driven away. The American urban street is utterly robbed, 'as though illusions of reality were stripped'. The built environment and its inhabitants are now left 'naked' and exposed. Even the buildings now seem to contemplate their mortality, 'the mystery charged with fearful clarity'. It all, he reminds us in the final line, has happened so quickly, in a single day, their minds 'racing' through the afternoon and 'on to the unhealing evening', when there will no longer be gentle dappled light through the trees and branches. Without the elms, the sun sets on 'insolent' death.[20]

From Wales, Gillian Clarke's poem 'Cardiff Elms' echoes Williams'. With the city bereft of its elms, 'Walls square up to the sky / without the company of leaves / or the town life of birds.' The elms, in memory, are presented as a sacred space within the profanity of urban life, the arches of their trunks forming a 'cloister' and their foliage, which she describes in terms of lace and softness, contrasting sharply with 'the parks / and pillars of a civic architecture'. She also reminds us that their spiritual role in our lives has long pre-dated the institutions of secular urbanity and all those city buildings. The trees were 'older and taller than all of it' and, by implications, better for us too. The poem ends uncompromisingly, witnessing a pile of dead trees, as if the loss of the elms is a symbol of the wider tragedy of our exploitation of the earth: 'elmwood, the start / of some terrible undoing'.[21]

The way that the loss of a tree reminds us of our own short lives is perfectly expressed by George Dillon in 'The Dead Elm on the Hilltop'. 'I have thought this tree could not die till I die,' he says, but it is struck and killed by lightning. Now 'summer stands no more in singing green'. Many of these poems use music and song to express the beauty of the elms, as if their elegance is melodic. One of the essences of poetry is it helps us to be a bit synaesthetic: different senses overlap and we smell colours or taste sounds. As we read words off the page with our eyes, we hear them in our heads, and all the patterning of sounds in poetic writing, whether through rhyme or assonance or consonance, accentuate that visual–aural confluence. The final couplet of Dillon's poem chimes with that earlier 'green' and presents us with the ultimate image of the viciousness of the death of the elm, as 'autumn, returning like a murderer to the scene, / Finds nothing left to kill'.[22]

One of my favourite elm poems is by the Romantic poet Thomas Hood, whose own life was short. Like other poets of the period, notably John Keats, he presaged his own death in his writing. In 'The Elm Tree' he makes brilliant use of ballad metre and an almost incantatory repetition of certain key lines to create an increasingly spooky atmosphere. The poet begins by walking down 'a shady Avenue, / where lofty Elms abound' and as the tale progresses he invokes a whole forest's worth of other trees – oak, maple, plane, ash, fir, beech, birch, aspen and pine – but all of them are upstaged by a particular elm that makes 'a sad and solemn sound, / That sometimes murmur'd overhead, / And sometimes underground'. He explores all possible sources of the noise – wind, sea, animals or birds – but it's a calm day, miles from the coast and no creature seems responsible, so he realises that the tree itself is sighing, moaning, making a 'hollow, hollow, hollow sound' and it does so until it is felled by a jobbing 'Woodman', whereupon there is an unearthly hush: 'No Zephyr stirs: the ear may catch / The smallest insect-hum'. Birds and animals sneak off in fear as into this silence stalks the spectre of Death, and for fourteen stanzas he stands beside the fallen tree to declare revenge on men, because 'When Elm or Oak / Have felt the stroke, / My turn it is to fell!' Death's speech is about how he levels society – from kings to tramps, 'One doom shall overwhelm!' – and also makes clear that an elm coffin is the inevitable destination of most people, particularly those without much wealth: 'he, who never knew a home, / Shall find it in the Elm!' After proclaiming the inevitability of the end of life, Death exits, and gradually the woodland creatures return, but the poet is left, 'sadden'd', with the sinking realisation – 'A secret, vague, prophetic gloom' – that the mystic elm is 'the fore-appointed Tree, / Within whose rugged bark / This warm and living frame shall find / Its narrow house and dark'. In other words, the fallen elm is destined to be his own coffin.[23]

Elms crop up a few times in William Wordsworth's vast outpouring of rural poetry, and they are firmly depressive in what is sometimes considered his first mature poem, 'The Ruined Cottage', a long, rambling tale within a tale, in which he meets a traveller who tells him a story. This character is first encountered stretched out beside the ruin of the title, 'beneath a shade / Of clustering elms that sprang from the same root . . . His eyes were shut; / The shadows of the breezy elms above / Dappled his face'. When he wakes, he relates the sad life of the woman, Margaret, whose house it was, as her family descended into poverty. Her husband abandoned her and her troubles continued with the loss of her children, the neglect of her garden and finally her own death. The traveller makes frequent mention of the 'lofty' elm trees as he explains the ruin's history of desolation. Wordsworth finds himself chilled by the tragic story he relates in 'that breezy shade', and has to step out from under the elms to be able to bear to hear the ending.[24]

Brad Leithauser paints another grim image in 'Dead Elms by a River', beginning with a late winter thaw and then a spring in which, unlike all the other riverside trees, the elms fail to come into leaf. From their tops 'thick with splinters' to their roots, under 'a frigid inch of water', they are visibly desolate and this misery infects everything else in the ecosystem, turning it hostile. In the second stanza birds, unseen, produce 'fibrous cries encircling', as if they were ropes strangling the trees, and in the next stanza other plants 'will close upon the elms', 'battling' with them. The elms are left in a perpetual, desiccated winter, 'a dry company' of 'ghosts'. I so clearly remember, from my own stream-valley childhood woods, those wintry, ghost-like forms among the lush summer foliage. As the poem progresses, the poor trees are gradually weakened and we are shown them disintegrating as 'the brittle / Outer branches are torn, / Stripped by wind and rain'. Gales rip down this river valley, 'peeling / Back their bark', until finally one succumbs to a storm, 'Snapped like a matchstick and lying / Tossed into the river', and it is clear this will be the fate of all of them. The poem ends with the bleakest of emotional messages. Rather than evincing any kind of sympathy, the elms are described as having 'airless hearts' that become increasingly hard, 'indurate', like diamonds, and ultimately, turning full circle to the winter from which the poem began, and which the elms have perpetuated throughout, the poem ends with the elms constricted into a 'core of ice'.[25]

Years ago, when I edited an anthology of tree poems, I was struck by just how many Japanese and Chinese poets write about pine trees. Modern American poets seem to undertake a similar rite of passage with elms. The Nobel laureate Louise Glück's poem 'Elms' takes three lines to get there but then tells us of her 'bitter sadness', caused by her concentrated gaze on the trees, which have shown her that 'the process that creates / the writhing, stationary tree / is torment'. No holds barred or subtlety here. She considers, and says she understands, the elms from a woodworker's point of view, as if being a poet is a kind of whittling or carving. Her conclusion is that elms will give only 'twisted forms', and perhaps that sums up the experience of so many poets, who can only muster something utterly grim in response to the trees.[26]

ACCEPTANCE

Like Leithauser, Paul Smyth uses ice as the central image for elm in his poem 'An Elm Leaf', a portrait of a single scrap of foliage in a New England November, but his is not a hopeless freeze. The light has the 'chill / Brilliance of quartz' and the leaf 'clings and shivers' above a pond. It is a 'brittle tongue' with a 'prophetic voice', to which we must listen to remain fully in this world, and not succumb to the 'velvet coffin' of abstractions and contemplation. The dead leaf contains within it the year that has passed, from its spring bud and its 'weighty green' summer growth, and now it is 'exhausted' and 'stunned' by frost, but it is not finished yet. The poet listens to its 'stubborn voice' and imagines its future. First it will 'skitter' on ice and then cause its surface to melt, before sinking in and being frozen over, becoming fossilised, far from lost and with all its memories of life retained in its form. 'What else but this,' the poet asks, 'is poetry or prayer?' What seems clear to the poet is that the voice of the elm will continue to be heard, long beyond its death.[27]

Andrew Young also seems to find acceptance through close scrutiny in 'The Elm Beetle'. He spent a long time looking at their 'strange hieroglyphics' that they had bored into elm wood, but rather than demonising them, the poem elevates them to 'small priests', the scratchings of which tell of spiritual journeys and adventures. The metaphor of the beetles as creators of an Egyptian script is extended into an evocation of Osiris, the Egyptian god of fertility and the afterlife. The poet emerges from his contemplation as if from a tomb, and the world he re-enters seems to be one of rural peacefulness.[28]

Aldous Huxley is best known for his mould-breaking novel *Brave New World* and his exploration in *The Doors of Perception* of the use of hallucinogenic drugs, essential reading for every teenager with a mind for experimental mind expansion. But he was also a poet and editor of poetry, and he wrote a note-perfect and prescient sonnet about elms. 'The Elms' presents us with the tree's spring show, the 'billows of emerald foam' that they present in April. In the early spring when all other trees are still dormant, the elms are miraculously beautiful, 'Seeming less trees than things beatified'. There is a mystical dimension to the trees, which he sees as having 'come from the world of thought', as if our vision is actually a form of conjuring and the trees, so perfect in their form and the light, 'powdery' green effect of their early seeding, floated into the world out of our ideas of trees like mind-expanding drugs: from their wonder he has been induced to wondering. All good sonnets set out in one direction and then turn; the Italians call this change in trajectory *'volta'*. Huxley's sonnet adopts the common convention of having the *volta* after the eighth line, with a sestet that shifts from wonder to warning. These final six lines are full of foreboding about the mortality of the tree. In a few deft lines they become clothed in their full summer foliage, a 'mountain-mass of clotted greenery', then face death: 'Their immaterial season quickly past, / They grow opaque, and therefore needs must die.' Once more the physical reality of the trees gives way to the ideas that they generate and their symbolic link to death brings the poem to its grim conclusion. And yet, the final line seems to show a kind of acceptance of the end of life: 'Since every earth to earth returns at last.'[29]

The older Wordsworth achieves a sanguine approach to elms, and in Book 6 of *The Prelude,* he is reminiscing of his happy days walking in Revolutionary France, where he finds 'benevolence and blessedness' in elm-lined avenues:

> Where elms for many and many a league in files,
> With their thin umbrage, on the stately roads
> Of that great kingdom, rustled o'er our heads,
> For ever near us as we paced along:
> 'Twas sweet at such a time, with such delights
> On every side, in prime of youthful strength,
> To feed a Poet's tender melancholy
> And fond conceit of sadness, to the noise
> And gentle undulations which they made.

These elms are clearly symbols of independence and freedom, as we have seen throughout revolutionary America and Europe. Wordsworth manages to weave in a reference to their classical association with vines, the ancient symbol of romantic love, which is appropriate as he fell madly in love with a French woman on that journey and had a daughter by her who he never got to meet, having had to flee the country to escape the war. This romance resonates in the lines that follow:

> Unhoused, beneath the evening star we saw
> Dances of liberty, and, in late hours
> Of darkness, dances in the open air.
> Among the vine-clad hills of Burgundy[30]

Lord Byron is much more straightforward in his love of a particular elm, under which he wishes to be buried in the churchyard of Harrow on the Hill. His 'Lines Written Beneath an Elm in the Churchyard' is full of adoration for the 'drooping elm' under which he has often lain and envisaged his own grave 'beneath this mantling shade'. It's a sonorous elm, 'whose hoary branches sigh' and are later 'moaning to the blast . . . and seem to whisper' that he should take his last farewells while he still can. He buried his daughter under this very tree.[31]

The most famous graveyard poem of all, and apparently the one most learned by heart, is Thomas Gray's 'Elegy Written in a Country Churchyard', another profound piece of levelling rhetoric, in which the poet ruminates on how the people buried 'beneath those rugged elms' might have been leaders, famous poets, artists or otherwise significant people had they not been so poor.[32]

In 'The Elm in Home-Ground', the shadow of the 'green elm' enables William Barnes to recall people from his earlier days. The tree's 'sheltering shroud' acts as a gateway to the author's memories.[33]

Robert Bridges finds not others but his better self under 'an old Giant' in 'The Great Elm', which he describes as:

> A towering Elm that stood alone,
> Last of an ancient rank,

> And had great bark roots out-thrown
> To buttress up the bank[34]

I'm immediately reminded of the elms that shored up the riverbank in my home village. The portrait continues from the tree's 'rough trunk' to its canopy's 'caves of leafy screen'. Under this tree he encounters his 'other half-self', an elf spirit that makes him feel complete, granting him a comforting sense that he is 'so strong / And nobler' than he had given himself credit for. He knows this spirit is fickle and hard to find, and wishes to 'hold him fast by magic rhyme / Forever to that tree'.

Matthew Arnold, the nineteenth-century English poet, had a soft spot for elms, which turn up in numerous of his poems, but most famously in 'Thyrsis', his long, Keatsian, elegiac ode to Arthur Hugh Clough, his close friend and fellow poet. At university they used to ramble in Oxfordshire and pledged under a particular elm tree an allegiance to a fabled Oxford academic who went feral and became 'the Gipsy-Scholar'. When in his grief he finds the tree is still standing, 'The tree! The tree!' he exclaims as it brings his old friendship back to him. Behind it, there is a sunset, of course, and its beauty and the steadfastness of the tree give him comfort: 'Despair I will not, while I yet descry . . . That lonely tree against the western sky.'[35]

We can always depend on William Carlos Williams to give a fresh perspective, and elms are no exception. Where everyone else focuses on the visual or oral dimension of the tree, he gives us an olfactory experience of an otherwise silhouetted tree in his 'Love Song', which scatters 'little loaves / of sweet smells' from its 'black branches'.[36] In his lines he captures perfectly the honeyed-dough smell of sunshine through elm leaves and the way a moment of bliss, or safety, can be remembered even in a dark time.

In 'The Elm's Home', a long poem by William Heyen, we get a suburban landscape minus a well-loved elm tree, which used to stand in one corner of the garden. While the poet is bereft, missing the tree, he has achieved a point in the grieving process where his love and memories transcend his sadness, and the poem is full of images of light. The tree's 'stump is a candle / of slow decay' on which grow 'clusters of tiny noctilucent mushrooms'. Just as a dead loved one can remain a presence, he can still see the tree, which whispers *'welcome home'* to him.[37] The title's lovely ambiguity is clear – this garden is the elm's home but also the elm itself is the poet's home. Contemplating the beauty of its decaying remains, he sees down 'into the ground, into the elm's dead / luminous roots'. The darkness of soil is transformed by these roots, which he describes as 'the branches of heaven / under the earth'. How literally uplifting!

I could go on forever, finding new elm poems and their litanies of grief, but as elms make multiple appearances in the poetry of Wilfrid Wilson Gibson, let's give him the last word. There's a lovely one called simply 'Trees' in which he sits under an elm by a fire, which 'half lit the cavernous mystery / Of the over-arching elm'. But for completeness we can't better his long poem 'The Elm', in which he narrates a lifelong experience of the tree, taking us through each of the stages of grief. The poem is stimulated by the falling of the 'brave' tree in strong wind, in which it 'snapped like a match-stick'. As if in denial of its loss, he falls to reminiscing about how it has been there since he had to stand on tiptoe to look at it out of the window, and relates its character in different seasons, from 'Soaring with all its countless leaves / In their first glory of fresh green / Like a big flame' to 'stark against the red / Of winter sunrise'. Repeatedly he describes it as 'soaring', until the night his wife died, when he looked out at it in the darkness and saw its 'great funeral-plume of black'. Now his adoration of the tree is extinguished and denial ends. For a while he is angry at the elm, then hates it as he bargains with it for his wife, but that sense wears off eventually until the night it is blown down, when 'he grieved to see it fallen low, / With almost every branch and bough / Smashed into splinters'. The grief is heavy, as if he is 'dragged . . . down' with the tree, but in the final stanza of the poem he emerges from that depression and decides 'he could make its trunk his seat'. An old man now, he hobbles out and makes his peace with the tree: 'He'd trust / His body to that wood.' In the last few lines, in the form of his grandson, the future beckons. [38]

Bill and I live on his croft, eleven hectares of woodland, bog and crags, in a mix of little shelters dotted about the place: there are three small caravans, an architect-designed-and-built studio where I work, plus numerous sheds, the largest of which we call 'The Great Hall' on account of its lofty roof space and beams. It's not, unfortunately, made of elm. Nor is the much more modest shed known as 'The Library', where I try to find space for my books. Shelves already cover the windows on one wall and will shortly claim one of the two remaining windows too, as every other centimetre is occupied by books. I have a shelf of elm books but if I kept all the books that mention elms, I would have a wall of them, including a whole shelf of fiction. I can't do all these books justice here, but the following are a few favourites.

David Copperfield remembers the elm trees of his youth with fondness, and they also appear in other Charles Dickens novels, including *Martin Chuzzlewit* and *The Mystery of Edwin Drood*. As such an archetypical English tree it's no surprise to find them in several D. H. Lawrence works, including his first novel, *The White Peacock*, *Sons and Lovers* and various short stories. There are graveyard elms in Bram Stoker's *Dracula*, and several in George Orwell books, including *Keep the Aspidistra Flying*, *1984* and *Homage to Catalonia*. Virginia Woolf has elms with stars and starlings in her novel *To the Lighthouse*, set in Skye, but actually based on a lighthouse in St Ives in Cornwall, and although Richens, in his amazing survey of elms in literature, accuses her of being 'notoriously botanically inaccurate', there are in fact elms in both places. This inclusion of elms in novels isn't just a British thing. The hero of Italo Calvino's *The Baron in the Trees* spends the book living and moving between the canopies of numerous species, including elms. Action goes on beneath American elms in Harper Lee's *To Kill a Mockingbird* and they're there as well in the New England settings of Vladimir Nabokov's *Lolita*. Dostoevsky has elms in his last novel, *The Brothers Karamazov*. Elm is repeatedly used as a symbol or augury of something significant – usually, you guessed it, death.

A small elm appears in a short story by Leo Tolstoy, 'The Raid', in which the protagonist seeks to understand the nature of courage, particularly in war. A civilian, he wants to join a group of military men as they go into action, in order to try to witness bravery in battle. The captain, Pavel Ivanovich Khlopov, reluctantly takes him along, travelling through a steep ravine, where 'the grey and whitish rock, the yellowish-green moss, the dew-covered bushes of Christ's Thorn, dogberry, and dwarf elm, appeared extraordinarily distinct and salient in the golden morning light'. The elm, as noted by Richens, is usually placed in fiction as tall, towering, lofty or otherwise vast and imposing, but Tolstoy's elm is dwarf in this quiet study into the soldiers risking their lives in battle. [39]

A more striking example of elm in Russian fiction is 'The Scarlet Flower' by Vsevolod Garshin, whose tragically brief life – he died aged just thirty, in 1888 – was marred by mental illness. The story, partly autobiographical, tells of an unnamed patient in a psychiatric hospital. The patient is only occasionally lucid and the story shows how what may seem incredible to an observer – the patient deems a red poppy to be a lethal threat – may be so real to the mentally ill man. The elm tree stands as a kind of generator of this imaginative power. The patient believes himself to have extraordinary abilities: 'He could read the thoughts of other men; he saw in objects their own history; the large elm tree in the hospital grounds told him whole legends from the past.' In this tale, the elm contains a vast and mystical cultural memory.[40]

In the Irish thriller *The Wych Elm* by Tana French, a wych elm is revealed to have been harbouring for the past decade, in a big hollow in its trunk, the victim of a murder. The elm tree once again stands in as a repository of memories, quietly holding its secret until the skull is found in the elm tree's hole. The elm tree becomes increasingly sinister, casting its shadow over the garden:

> It was one of the biggest trees in the garden, and the best for climbing: a great misshapen grey-brown bole, maybe five feet across, lumpy with rough bosses that made perfect handholds and footholds to the point where, seven or eight feet up, it split into thick branches heavy with huge green leaves. It was the same one I'd broken my ankle jumping out of, when I was a kid; with a horrible leap of my skin I realised that [the skull] could have been in there the whole time, I could have been just inches away from it.

But the tree has been holding quietly onto its secret, as one character points out: 'We haven't spent the last ten years hearing skeleton fingers scrabbling inside the wych elm whenever we walked past it.' The police insist the tree is felled in order to be thoroughly searched. The result is 'carnage – huge branches strewn across the grass, sawdust flying'. Drama and dialogue are cut through by 'bursts of noise from the chainsaw' and 'unpredictable roars from the chainsaw making me jump every time'. Afterwards, the garden 'looked like some old battlefield, World War I, flung heaps of dirt and lopsided holes' and 'the jagged crater – shockingly wide and deep – where the wych elm had been'. The tree's secret changes everything; the space it leaves is spooky and it remains an indefatigable symbol of death.[41]

There's an echo here with Thomas Hardy's *The Woodlanders*, in which John South is convinced that the elm tree outside his house is going to fall and squash him:

> I could bear up, I know I could, if it were not for the tree – yes, the tree 'tis that's killing me. There he stands, threatening my life every minute that the wind do blow. He'll come down upon us, and squat us dead.⁴²

The elm is first pruned of its lower branches to reduce its susceptibility to wind, but this makes South even more frightened because it makes the tree seem even taller. The tree is felled to ease South's 'craze about the elm', but instead South's 'whole system seemed paralysed by amazement . . . He lingered through the day, and died that evening as the sun went down. "Damned if my remedy hasn't killed him," murmured the doctor.'

In an otherwise perfectly 'realist' novel, this very strange, almost supernatural, event has been much chewed over by literary critics. Most agree that Hardy is deliberately setting up an identification between South and the elm, what Fiona Stafford calls 'the complicated intimacy of people and trees', though she also refers to it as a 'parable of neurotic psychology'.⁴³ John Miller describes it as 'a symbolic evocation of human-forest symbiosis', with South and the elm in a state of co-dependency: 'its wellbeing is his own'. But then he adds, 'It is equally possible to read the scene not in terms of symbiosis but as an expression of arboreal revenge.'⁴⁴ This elm tree is being granted a level of retributive agency that we don't normally associate with plants, and Maxwell Sater believes that Hardy is doing this 'precisely in order to frustrate and disappoint our imagined powers of ecological reasoning', as a way of showing that the woodlanders of the book's title possess knowledge, and have ways of reasoning, that most other people don't have.⁴⁵ South's daughter, Marty, lives a life completely entwined with and dependent upon the woodland, rooted into it as if a tree herself. She describes her father's relationship with the elm as follows:

> The shape of it seems to haunt him like an evil spirit. He says that it is exactly his own age, that it has got human sense, and sprouted up when he was born on purpose to rule him, and keep him as its slave. Others have been like it afore in Hintock.

There is nothing unusual to her in the tree having agency and being capable of enslaving a human being. Just as Aldo Leopold advocated 'thinking like a mountain', Hardy, Sater claims, wants us to see the Souths and other characters as thinking like the woodlands of which they are a part.

Another elm tree presence with fatal consequences is in American playwright and Nobel laureate Eugene O'Neill's 1924 play *Desire Under the Elms*. The play's action is entirely within and beside a house over which 'two enormous elms . . . bend their trailing branches'. Their presences 'brood oppressively' through the play, which begins unhappily and ends in tragedy, with the trees representing what O'Neill describes as a 'sinister maternity'. Like Hardy's elm, they are paradoxical in both affirming and denying life: 'They appear to protect and at the same time subdue.' Also like Hardy's elm, in the opening stage directions of the play, they are given explicitly human-like character: 'They have developed from their intimate contact with the life of man in the house an appalling humaneness . . .' The one female character in the play, Abbie, explicitly compares herself to the elms, claiming that the sunshine 'makes ye grow bigger – like a tree – like them elums'. Abbie is there to lay claim to ownership of the house but it is the explicitly – malevolently yet miserably – female elms that are really dominant over it, as O'Neill has made clear in those opening stage directions: 'They are like exhausted women resting their sagging breasts and hands and hair on its roof, and when it rains their tears trickle down monotonously and rot on the shingles.' No mere human woman stands a chance of taking control while they are around.[46]

This is the polar opposite of E. M. Forster's wych elm in *Howards End*, which broods similarly over the house of that name, but in this case it is only women who stand any chance in the building. The men of the Wilcox family, who are so rich they have many houses to choose from, all loathe the place but are drawn back to it again and again due to the magnetic attraction it has for one or other Mrs Wilcox, and even though he claims not to like the house at all, Henry Wilcox declares that he 'shouldn't want that fine wych-elm spoilt'. It is mentioned on the very first page of the novel, when Helen Schlegel writes to her sister about it, declaring, 'I quite love that tree already,' and the kiss that sets the story in motion 'under the column of the vast wych elm' must be the most famous kiss under an elm tree. When Margaret, the second Mrs Wilcox of the story, first sees the old wych elm, she is stunned:

> No report had prepared her for its peculiar glory. It was neither warrior, nor lover, nor god . . . It was a comrade, bending over the house, strength and adventure in its roots, but in its utmost fingers tenderness, and the girth, that a

dozen men could not have spanned, became in the end evanescent, till pale bud clusters seemed to float in the air. It was a comrade.[47]

This and other elms have ominous power in the novel, reappearing repeatedly through the story: at a funeral, there is a woodsman up an elm, pollarding it, which causes great disgruntlement. Helen says at one point, 'Under that wych elm – honestly, I see little happiness ahead,' but on the whole it seems to have a benign influence. It has pigs' teeth stuck into the bark, and is claimed to be able to cure toothache. It's the scene of a key confession at the climax of the novel, at which point even its rustle is transcendental:

> It had made music before they were born, and would continue after their deaths, but its song was of the moment. The moment had passed. The tree rustled again. Their senses were sharpened, and they seemed to apprehend life. Life passed. The tree rustled again.

In the final few pages of the novel, Margaret considers that 'every westerly gale might blow the wych-elm down and bring the end of all things'. Far more than a mere tree, the elm is at the very least a witness of all the key moments of the players' lives, and is finally given the role of conductor of the score. But what chimes for me most of all about the tree at Howards End is that its 'message was not of eternity, but of hope on this side of the grave'. The elm is, ultimately, an uplifting presence.

Visual art, like the natural world, makes contact in ways that bypass the rational, verbal mind – as someone who spends a lot of time wrestling with words, I find this intensely liberating. Some of my happiest times have been when collaborating with visual artists, such as for 'Trees Meet Sea' in the Sawyer Gallery at Inverewe Garden in Wester Ross. This was an exhibition of artworks in various media, by fourteen artists, paired with my poems – for half I'd written a poem in response to the art, and for the other half the artists had responded to one of my poems. As the gallery filled with panels of my words nestling close to the painting, textile, sculpture, video and ceramic, I was overwhelmed with excitement. The walls worked at first like illustrated pages, but as we shuffled the frames and words something wordless and vibrant took over, and the room became a humming space as its three, or perhaps more than three, dimensions enfolded the complexity of it all. The pieces became a whole and danced with each other and with us, the observers, or participants in the show. The choreography was a team effort with curator Adrian Hollister, artists Lynn Bennett-Mackenzie and James Hawkins, and all-round art wonder Flick Hawkins, and when we moved the exhibition to Dundee Botanic Garden, the magic happened all over again, thanks this time to curator Kevin Frediani, with James and Flick once more bringing the inert pieces to life.

I'd love to see an exhibition of elm art, a gathering of some of the mighty works that celebrate and respond to these marvellous trees, and wonder what kind of mystical forest would result. It is simply not possible to condense into the black and white of alphabetical script the art made from elms over the centuries, so instead I'll attempt to curate a small, imaginary exhibition of some of the pieces that have most moved me.

In the most prominent, central position in the gallery I have to place John Constable's *Study of the Trunk of an Elm Tree*, from the Victoria and Albert collection in London.[48] It's so vivid as to be almost believably a photograph or even an actual tree. It's from the perspective that you see when you have walked up to an elm: from just where it sprouts out of the grassy ground, its mossy trunk adamant and unquestionably strong, steady and firm. The painting goes as far as its lowest branches, well above head height. They're in full leaf, but as the title suggests, the focus is on the trunk, not the transience of foliage. The bough that is cut off at the top reminds us that we could crane our necks up into the canopy, or look through to the nondescript vegetation beyond or the sky, but why would we? This chevron-patterned bark, this oh-so-huggable tree, is all we need to see.

On one long wall of my imaginary gallery I have etchings and drawings. Pride of place goes to as many of Ian Westacott's extraordinary chronicle of the great elms of the Highlands as I could fit.[49] With monocolour marks, he conveys the full character, the total majesty, the age, fragility and decrepitude or the strength, buoyancy and lightness of living trees. His works contain the merest suggestion of a world beyond the tree – he gives the trees all the air and space of the white sheet and all the wind and gales and gentleness of his lines, and the result is that the trees are real characters. Somehow the starkness of his black-and-white images reveals all the colour and sound and drama that the trees embody.

At the end of Ian's wall, I'd place *The Waterloo Elm*, an almost leafless and branchless tree portrayed by Anna Children in 1818. This elm stood on the site of the Battle of Waterloo, where the British commander was supposed to have been positioned. Such was the adulation for him after victory that the tree was ravaged by people seeking souvenirs, and when Anna visited the site, only the bare trunk remained, which she captured in her shrieking shock of a drawing. The tree was felled and acquired by the artist's father, who had a chair made from it, and in 1821 the chair was gifted to King George IV, along with the drawing, which remains in the Royal Collection.[50]

I'd use the other long wall of my elm gallery for photographs, starting with those by William Henry Fox Talbot, the Englishman who is credited with inventing the process of photography almost simultaneously with Frenchman Louis Daguerre in 1839. Talbot's *Elm Tree in Winter, Lacock Abbey* is a late-winter tree, about to burst into flower – its burgeoning buds are visibly blurring the otherwise stark lines of the twigs.[51] It's a tall, supremely elegant elm, its trunk appropriately just skew to the vertical, as if the prevailing wind comes from the left side of the picture, so its profile has the asymmetrical shape of an individual elm leaf. The rounded crown is open to the sky. It is a rare, perfect portrait of a whole tree.

Just as Talbot used the latest technical breakthroughs of his era to showcase elms, so did Francis Principe-Gillespie in his microphotographs of elm's most intimate secrets, so I include his close-ups of the miraculous details of elm flowers. I'm sure everyone will be amazed to see their tiny pollen-filled anthers like big leather purses bursting with grains of cous-cous, and the stigma looking for all the world like a bright pink sea-slug.[52]

The rest of the photography wall is taken up by Chris Puddephatt's gallery of treasures. While I've been mooching about scribbling in my notebook under trees in Assynt, Chris has been carrying out a far more methodical and comprehensive pilgrimage around all of our lovely elms, and the result is a gallery that has blossomed into something that goes way beyond a botanical record. He has captured the elms in all their moods, in all weathers, all seasons, all times of day and from all angles, from wide-angled placements in our spectacular landscape to close-ups of the myriad epiphytic life forms that bedeck them. They're available online and I can offer no better place to see how I got to be as passionate about our elms as I am.[53] I'd organise them by season, starting in winter – with monochrome snow and blanched landscapes with stark silhouettes – and run through the lushness of spring and summer to end with all the gorgeous colours of an Assynt autumn.

The final wall is graced by George Clausen's *Elm Tree in Spring,* bedecked with golden samaras.[54] This would help to soothe us from Edvard Munch's *Elm Forest in Spring,* the vibrant, almost garish blur of colours of which makes me feel a bit queasy, as if a landslide is underway. It's one of many elm forest paintings by Munch, in which the trees are so vividly alive they seem about to step out of the picture.[55] I'd also include *Elm Trees* by Philip Wilson Steer, who conveys in a few lovely green blotches the paradoxically princely presence yet unpretentiousness of the tree, always just that bit lopsided, elegant at first glance but a bit shambolic when you look more closely.[56] The paintings wall would be completed by a classic piece of English pastoral, Alfred East's *Autumn in Gloucestershire,* with its huge elms scattering their leaves, inviting us to scuff among them.[57]

Right in the centre of the gallery floor would be Henry Moore's abstract *Hole and Lump* sculpture.[58] What better way to include the presence of an elm than in the polished wood? The lustre of its grain shines and the artist's form follows and enhances the curves and protrusions made during the long weathered life of the tree, the grand solidity of the piece echoing its rooted origins. I'd also want Barbara Hepworth's *Pelagos*, its form reminiscent of Orpheus' lyre, and her *Dryad*, bringing a forest spirit into the space.[59, 60] A piece of the Preston elm, gilded by Elpida Hadzi-Vasileva, on its elm wood plinth, completes the sculpture section.[61]

The gallery itself is an elm barn, of course, and I'll leave space for you to bring your own favourite pieces for display.

Bella's Elm

The Wych Elm novel by Tana French echoes an equally gruesome true mystery.[62] In Hagley, Worcestershire, there stood a huge, unquestionably ugly, wych elm tree. It had been much pollarded over probably hundreds of years and the pictures show a great, lumpy barrel of a trunk topped with a scrubby, shrubby mass of little branches. The trunk was hollow, and on 18 April 1943, a woman's body was found in it. Or, at least, the remains of a body. By the time of its discovery, only a crooked-toothed skeleton and some remains of personal effects survived – a cheap ring, a woollen belted cardigan, crepe-soled shoes.

Forensic experts believed that she had been dead at least two years, but no one knew who she was. Then, about six months later, 'Who put Bella in the Wych Elm?' was painted on a brick wall in town, and 'Who put Bella in the Witch Elm?' has appeared on the tall stone obelisk on a nearby hill. Bella is a mystery.

Theories abound, of course. Was she a Nazi spy? Was she the victim of a satanic ritual? Was she just a poor woman, victim of a murder, body stuffed in a tree? The lack of answers has generated so much speculation that researching it can feel like disappearing down the black hole in the heart of that hollow tree.

Chapter Ten

Life: Lessons from the Elms

We live in a rapidly warming world. Over recent decades, we've seen rising summer temperatures, changes in rainfall distribution and more frequent extreme weather events like storms, floods, droughts and fires, resulting in existential threats to populations of low-lying islands, Inuit hunters and polar bears, but also more mundane and obvious signs like transport disruption. Our holiday choices, daily commute, home heating, gadgets and even the food we eat can contribute to global warning. This ubiquity and complexity makes climate change what philosopher Timothy Morton calls a 'hyperobject', impossible to perceive in its entirety from any perspective. Many of us find ourselves frozen, as if in headlights, by the need for so much to be addressed, the difficulty of prioritising or choosing which action to take, knowing our own contribution will be indetectable and thus feeling it may be pointless.

In the year leading up to the death of the Beauly Elm, I turned my academic research to how, in my local area, we might achieve net zero carbon emissions. I scored a grant and began asking people how they felt about climate change. I was horrified by the extent and depth of young people's fear, dread, anxiety, panic, anger, fury and just plain sadness. Several months into the project, I was drowning in hopelessness.

Elm trees lifted me out of that.

Elms are threatened by global warming as much as everything else: those in places that are getting dryer will suffer from drought; fires, floods and storms will take their toll; and disease risks are exacerbated by warming. As I have already described, over the course of my lifetime, isotherms – the lines on the map joining places with similar temperatures – have been moving steadily northwards, and each shift opens up a new population of elms to elm bark beetles. Disease control is becoming more difficult due to the longer hot seasons: in Brighton, for example, there used to be two hatchings of beetles each year but now there are usually three.

When I was a child, my grandfather gave me a big, white teddy bear to protect me, and polar bears have always felt like my totem animal. I love them with a deep passion and my most sacred memories are of my encounters with them. Seeing a piteously thin bear eating seaweed in Svalbard made my heart break. As did the emaciated man on a tiny island in the Bay of Bengal, who I watched carrying mud to shore up his field boundary in an attempt to protect his crops from the rising sea. I felt guilty that climate change will merely make our cool Scottish climate somewhat warmer, our summers a bit longer and our winters even wetter while tragedy impacts distant places like the Arctic and flat, tropical islands. Elms have connected climate change to my own life: from the landslide that trashed my Northumberland childhood playground and companion trees, to my student walks across the parks of Edinburgh, to happy times in Beauly on first moving to the Highlands and now the place I've rooted in for twenty-five years and will remain in, I hope, until I die – all of them show me elm trees toppling as the world gradually warms. My life has followed a northwards track at exactly the same pace as the isotherm that enables elm bark beetles to spread further. For the past fifty years, the beetles have loved the same new places as me at the same times, and the lovely elm trees have paid the price. Climate change is no longer just a huge, planetary scale matter, but an effect that I have directly experienced. Instead of being an existential threat against which I am powerless, climate change has become a danger to an immediate friend who I must try to protect. Through this personal connection the elm has empowered me to act, and encourage others to follow suit.

There is a crucial and simple thing that we can do to address the challenges facing elms, and it is also one of the best actions we can take in response to climate change: plant them and encourage natural regeneration where there are existing seed sources.

In the past, here in Scotland, the powers that be offered rewards to those who planned significant increases in elm populations. In the nineteenth century, the Highland Society of Scotland announced: 'To the proprietor in Scotland who, from 1st October 1826, to 1st October 1829, shall have planted the greatest number, not less than ten thousand, of ash or Scots elms, and effectively fenced the same, in order to raise timber – a piece of plate of the value of twenty sovereigns, with a suitable inscription.' This challenge rewarded those already busy propagating and planting trees, but the prestige of the inscribed plate encouraged wealthy landowners to take part too, though it would have required a huge nursery and planting effort. It's high time that we had a similar reward for the planting of elms in the Highlands – not only to protect the elm but as a step in the fight against climate change too.

Despite national afforestation targets and the forestry industry's efforts, elms have not been planted in significant numbers over past decades. It isn't an attractive prospect for commercial forestry and since the arrival of Dutch elm disease, most land managers have given up on elm. But we need to keep planting elms. They are part of a big, powerful whole and are vital to many ecosystems. Elms, like us, interact in complex and interesting ways with other species, all stitched together into the rich and multicoloured tapestry of life. To understand what might happen to elms in the future, we need to understand everything from blackbirds to bark beetles, from owls to *Ophiostoma novo-ulmi*, the Dutch elm disease fungus itself. We can't appreciate elms or ourselves completely without acknowledging our ecosystems, and that sense of connection, of being part of a living system, is profoundly heart-warming.

Engaging with elms has immersed me in what another eco-philosopher, David Abram, terms the 'more-than-human world'.[1] The feeling of being a thread in the great net of life brings with it the sense that if I wriggle, my vibrations will travel in myriad directions – we're all in this together.

Elm trees, breathing in carbon dioxide and breathing out oxygen, are already busy doing their bit for climate change.

They're also demonstrating resilience. Since the last ice age, 10,000 years ago, elms thrived for 5,000 years, suffered a great decline and then recovered. To us, 5,000 years is hard to imagine, but it is only about six generations for elms living to their potential 800 years each. It's just a few generations since they suffered a population collapse, and in that short timeframe, to them, they have bounced right back, just as humans soon made up for the population crash of the great plague in the mediaeval period. While we wring our hands and write our elegies about the loss of elms, they are steadily working on their recovery, year by year, seed by seed. And while they're at it, carbon atom by carbon atom, they tackle our greenhouse gas emissions.

Another of elms' hopeful gifts is time. Time goes round and round: not only the hands of a clock, but the rhythms of the tides, the cycle of the moon, the seasons, and all of life – there's a good reason we talk about 'life cycles'. An elm tree demonstrates this clockwork circularity every year – impatient to bring an end to winter, they're one of the earliest trees off the starting blocks with buds, bringing early spring flowers, pollinated by March gales, setting seeds that ripen by the time summer races over the horizon, ready to be thrown to June breezes, germinating in the warm moist weather of midsummer so that by autumn there they are: miniature seedling trees. A new tree can already be established and growing independently by the time its parent tree begins next year's cycle. Although it will take a decade or more until that seedling is mature enough to fruit for itself, there is a beautiful completeness about elm's ability to reproduce itself within a planet's circuit around the sun. And there is extraordinary abundance in its seeding – thousands or tens of thousands of seeds on a single tree every year – so old trees such as the Beauly Elm and the Japanese zelkovas could have potentially produced millions of new trees. This is generosity and fecundity on a scale we mammals can't begin to understand.

While we flit between social media channels and are bombarded with breaking news, hurrying our way from minute to minute, reminded constantly of the need to act before it's too late, the elms stand still and grow. Trees experience time differently from us. Once we acknowledge that, time can feel different for us too.

Thinking like an elm tree has taught me the difference between which actions are urgent and which are important. Sowing seeds is an urgent matter, but only once a year. Most other climate action is important, but not urgent. Of course I believe that the sooner we address the changes to climate emissions that our society needs to make the better, but the transition to a fully sustainable way of being human is going to require thoughtful choices and activity for the rest of our lifetimes. Such change is systemic, deep and powerful, a bit like tree growth. So, climate action is vitally important, but it cannot be done in a head-spin of urgency.

Along with the rest of the environmental movement, I used the expression 'climate emergency', but over the past year of elm-watching, I've realised emergencies are events that require immediate, drastic, high-paced intervention, designed to bring the situation to an end. Climate action just isn't like that and I no longer believe the emergency metaphor is helpful: it implies that the problem will be short-lived, that experts will be able to handle it and that we should be in a state of heightened emotion, in 'fight-or-flight' mode, until help arrives. It makes many people so upset that they're understandably immobilised or frozen with fear, or too distressed to be rational. In reality we all need to engage deeply and long term, in cool, life-affirming ways. I believe recovery or healing is a better way of thinking about the issue: getting off our fossil fuel addiction, restoring our damaged relationship with the rest of the natural world and transforming to healthier ways of being.

Over the course of my lifetime, Dutch elm disease has shifted north, mile by mile, but it has taken 50 years since my childhood for the blight to reach Beauly, roughly 250 miles north. It is a slow wave, with a key limiting factor in its progression: the beetles' need for sufficiently warm weather to fly. The elms are our canaries in the pit, the signal of a life-threatening build-up of carbon dioxide. Just as the colliers of old knew watched for the canaries collapsing in their cages, when Dutch elm disease starts killing trees even here in the coolest north-west corner of Europe, that's a signal to step up and act.

Yet elm trees also teach us the strengths of inaction, because so often our attempts to smash one problem with a rushed, fire-fighting response end up causing damage to something else. In his book *The Insect Crisis*, Oliver Milman argues for an 'inaction plan', that what we need to do to help insects is in fact less of the things that hurt them, like our incessant tidying up or our use of toxic chemicals to prevent an explosion of one kind of insect (like crop pests) that end up causing massive harm to many more helpful ones (like pollinators). He says:

Reversing our destruction of insects can seem a complex challenge, involving the overhaul of a gargantuan agricultural machine, the evolution of cultural norms, and the decoupling of improved living standards from environmental annihilation . . . But if you squint a little, addressing the insect crisis can be viewed as surprisingly straightforward. In essence, we would just stop doing certain things. The mere absence of action, of letting things slide a little, could be enough.[2]

The same is true of climate change. Doing less may be more helpful than doing more.

Elms' have also taught me that hope can come from a shift in spatial scale. We humans like to think of ourselves as mobile, with our fast cars and ability to jet around the globe, but most of the time we are indoors, or in gardens or allotments no bigger than a stone's throw. We inhabit built environments designed for people rarely bigger than 2 metres tall, and we think about space from the perspective of a human body. But the kind of thinking that elms require of us is not about individuals, it's about the population as a whole at a landscape level – seeing the woods, not just the trees. This means not worrying so much about a particular tree, but instead considering the wider factors in the landscape that are detrimental to tree survival, such as heavy grazing or browsing pressure, flooding, drought or fire risk, and what we can do to help, such as identifying the best soils for new planting or culling or fencing out herbivores where natural regeneration can take place.

We may think that our mobility somehow makes us a 'higher life form' than a tree, but it often seems to me that stasis, the ability to stay in one place and know that place really deeply, is a much higher form of consciousness. Throughout this book, we've seen elm trees so often playing the role of holding memories in a significant place. Their longevity and deep-rootedness matter profoundly. During the COVID-19 lockdown, I became fascinated by barnacles, those little sea creatures who spend their youth journeying as what I call 'nauplius nomads' in the currents of the ocean, and then one spring day decide to stop, for good. Such is their certainty that they secrete an adhesive from their foreheads and literally cement themselves to their new and forever home. Then they set about their adult existence, joyfully – or so it seems, if you watch them closely and see their feathery little legs dancing and fluttering – in a perpetual and lasting handstand, not unlike the above-ground life of an earth-rooted tree. Trees, of course, stay where they are for their entire, long lives, and they can offer a place where we can stop and let something wise emerge by paying attention not to alarming extreme events but to the slow changes and unfolding natural processes around us. If elms are anything to go by, just standing still, breathing slowly and deeply, and calmly watching the trees grow might be the best action of all.

Over recent years we've seen across the UK a revival of enthusiasm for our native woods, and part of that is recognition that our western seaboard's natural vegetative cover is temperate rainforest, which was largely wiped out over past centuries and continues to be damaged through negligence or neglect. Guy Shrubsole's book *The Lost Rainforests of Britain*[3] has done a brilliant job of highlighting the value and magic of our western wet-woods, but that book has a crucial element missing from it: elm. Apart from one reference to Dutch elm disease, Shrubsole does not mention elms, presumably because so many have been lost in the southern areas where the book is focused. But if you come north to the rain-soaked land where I live and delve into one of our soggy glens, with its river torrenting through the bottom and huge epiphyte-festooned trees straddling out from ferny, mossy, craggy ravine-sides, it's not hard at all to see that elm is a key rainforest species.

Guy Shrubsole is by no means alone. Very often the impact of Dutch elm disease is wildly over-exaggerated, as if the only thing worth saying about them is that they are the victims of a fungus. Colin Tudge, in his book *The Secret Life of Trees*, begins with a brief description about their definitive contribution to English landscapes, and notes the value of their timber for chairs and wheelbarrows, but he then describes the impact of the disease as 'one of the most dramatic extinctions in historical times'.[4]

The misconception that elms are undergoing an extinction is widespread but premature. Fiona Stafford states that they have gone 'from essential to endangered to extinct', and 'now the few remaining places where elms survive to maturity have become sanctuaries'.[5] Although elms have long been associated with death and life thereafter, in these blighted times she suggests the tree has begun taking on other associations: 'The tree means lost strength, disappearance and nostalgia, or, perhaps even more poignant, it suggests youthful hope blasted.' I hope this book has managed to convince you otherwise. Yes, it's a tree that resonates with death and holds memories, but it's also resilient and can teach us to hope for a future life of liberty and peace. Barn-builder Robert Somerville sums it up nicely:

> In the context of our society's struggle against environmental loss and climate chaos, the symbolism of the elm is something grounding and earthy. It is one full of complexity and contradictions, but also a symbol of renewal, of survival against the odds and of life after tragedy. A tree of hope.[6]

However, negative views pervade. The world of forestry has given up on elms; their hedgerow habitats have been wiped from our farmlands; they have been replaced in cities by more resilient tree species; and most people have forgotten, if they ever really knew, how to identify them. As Robert Somerville says, 'Elms have become the forgotten tree, the tree we gave up on.' People have been led to believe that they have all been wiped out. But once you look, you spot them everywhere – on waste ground, along railway lines and of course in woodlands.

As the editors of a recent book about elms in Britain announce at the opening of its introduction, 'For too long, the elm has been spoken of in the past tense, whereas in fact it can be found in significant numbers throughout the British Isles . . . They spring up virtually everywhere the English elm has been historically prevalent and where they are allowed to do so.'[7] A tree that dies or is cut above ground can still have a perfectly viable root system, and if it resprouts it has at least a decade of growing before it reaches the height and scale when it's susceptible to the disease again, with big enough trunks for the beetles to bore into and create the brood chambers in which the fungi will grow.

On the day before my father's funeral, I went back to my childhood woodlands to see if there were any signs of the old elms. The elm stump where I spent my teenage years has long been swallowed up by ivy and brambles, but further along the top of the land-slipped riverbank, I saw that there are now clumps of relatively slim elms that have clearly all grown from other old rootstocks. By the size of them, they are maybe a decade or so old, so they look to have sprouted long since Dutch elm disease first arrived and appeared to kill the original trees. Perhaps there have been multiple waves of the disease through the woods, or perhaps the roots still harbour the fungus and it knocks the trunks back when they get too big. It's easy to mistake the resulting clumps for hazels, as they are adopting a similar niche in the woodland understorey, with bunches of stems springing from a single coppiced stool.

Up above are the big oaks, ashes and beeches that have taken up space left in the canopy by the elms that died back in the 1970s. I thought I would only find this bushy elm 'scrub' and no actual trees, but then, on the slope down to the stream that wends its way through the heart of the wood, I bumped into a perfect elm, its trunk still slim enough to hug right round, but only just, tall and straight and topped by a lofty, elegant crown. This must be a forty-or-so-year-old tree grown since Dutch elm disease culled the elders. It has competed successfully with the neighbouring ash and oak saplings that all grew in a thicket once the big elms fell, striving up towards the light, growing tall and straight.

The elms from my childhood woodlands have not all died out at all! They're not wiped out. Some at least are still there, very much alive.

Robert Somerville points out that in the fifty years since the height of the elm pandemic, there has been a 'catastrophic decline' not so much in elm trees but in 'our awareness and relationship with the tree. For fifty years, elm has been a secretive tree, and its timber a hidden treasure. Some elms are so large and tall that when you see them it really challenges your rationality. They really shouldn't be there. They should be dead.'[8]

What amazes me most about that huggable tree is that I've been walking past it for the past half-century. It has been hidden in plain sight. All my life long I assumed that a key element of my childhood had vanished. But on probably my last-ever visit to the village I grew up in, after the sale of the only house with which I have a family connection, for the funeral of the last member of my family to live there, I discovered that all along, what I thought was long gone was still there after all. The death of my father was a rebirth of the elms. His burial has restored something to me that I thought was irretrievably lost.

Dealing with the death of my father has involved unearthing, facing and attempting to process a lot of difficult things. His behaviour made me who I am and it is hard to disentangle the aspects of myself that I need in order to remain me while finally freeing myself of the detritus of childhood hurt. Elms make it look easy as they divest themselves of their leaves in autumn, but as winter storms shred twigs and down branches, I see that they too undergo their struggles in shedding parts of themselves. Trees may have their own kind of therapy. They certainly seem to offer it to those of us who need it.

When I started writing this book, Assynt was free of Dutch elm disease. Early in the project, however, I was told that elms were dying at Glencanisp. By the time I got there one had been felled but a couple of others nearby were already wilting. All five of the elms next to the big house seemed doomed. After that I started noticing other elm trees in the area that had bare branches or wilting leaves.

It cast a dark shadow. I'd been proclaiming myself to live in a 'zone of hope' but now Dutch elm disease had arrived and instead I seemed destined to chronicle the demise of the trees. While my heart sank, I redoubled my efforts to make sure that I would not simply stand by and watch them die, that I would draw local attention to the treasure that we have before we lost it. So I launched the Assynt Elm Project.

For several months after my father's death I wasn't fit for work, yet somehow if elms were involved, I seemed to function. No matter how incapacitated I felt, I continued to exchange emails with elm enthusiasts. The local community woodland group offered to help. I wrote an application for funding support for a local project about elms with our local woodland organisation, and it received support instantly. A national newspaper did a feature on our elms in their Sunday supplement, then a magazine followed. I was asked to contribute to blog posts and write an article. People were unbelievably keen to join in a celebration of our elms and to discuss how we could protect them. Chris Puddephatt set out on his photographic pilgrimage around the Assynt elms, and his stunning pictures helped to galvanise interest, sympathy and enthusiasm. Our local field club, our community tree nursery run by the Scottish Wildlife Trust, the Woodland Trust, the John Muir Trust, Scottish Forestry, Reforesting Scotland, the Alliance for Scotland's Rainforest, local landowners, the primary school, everyone and anyone got on board. The Royal Botanic Garden Edinburgh became enthusiastic supporters of the project and chose Assynt as a key site for their special elm breeding programme.

Dr Euan Bowditch, who runs the Highland Elm Project and is using DNA analysis to better understand elm's diversity and seeking any genetic resilience from disease, visited us to help locals identify elms. On a dreich, cold, wet December afternoon more people than I could have imagined in my wildest dreams turned out to see what they could do to help. Two dozen folk, happed up in their wet weather gear on a freezing-cold, wind-blasted afternoon, were enthusiastic to learn more from Euan about elms and the threat of disease, and listen to Chris explaining how to take good tree photos.

On a chilly January afternoon, my plans for an elm planting event came to fruition and forty-three people came to make portraits of elm seedlings and plant them. Out of a population of 1,000, that attendance is equivalent to half a million Londoners simultaneously taking to the streets to draw elm tree seedlings – if only! One of the trees we planted was special, so before we put it in the earth, we propped it up on a pile of stones and formed a 7-metre in circumference human circle around it so everyone could picture how big its trunk could grow. Max Coleman, our guest from the Royal Botanic Garden Edinburgh, had brought it with him. He told us proudly who exactly its parents are: its mother is a huge elm with a wide canopy, conveniently low to the ground in places, living in East Lothian, and its father is from Peebles. Of course, 'he' could just as easily have played the role of 'she' in the relationship, and vice versa. His pollen was collected in special bags attached to his branches for the purpose, transported to her and dusted carefully into her low-hanging flowers, also encased in a bag to prevent pollination from any other tree. When the seeds ripened they were gathered and sown and, once germinated, nurtured by Emma Beckinsale and her team in Edinburgh, then brought to Assynt to begin a new life. The two parents were chosen because they have survived multiple bursts of Dutch elm disease so their offspring is 'potentially resilient'. There is no guarantee, but we hope that whatever has enabled its parents to live through the disease will be passed on in their genes.

Since then we've done another elm art event, the school children have been learning about elm ecology and close to winter solstice we ceremonially made charcoal from the Glencanisp elm wood. Strangers keep contacting me to tell me how inspired they are by what we are doing about elms. All of them say that it gives them hope.

There have been intensive breeding efforts around the world for the past century, trying to identify and propagate elm trees that are resistant to or tolerant of Dutch elm disease. In the Netherlands, this has been going on since the very first years of identification of the disease and it continues, particularly with the production of cultivars that cross field elm (*Ulmus minor*) with Himalayan elm (*Ulmus wallichiana*), which doesn't suffer from the disease, presumably because it lives with a close relative of the fungus, *Ophiostoma himal-ulmi*.

In Italy there is further research looking into crosses between native elms and Siberian elm (*Ulmus pumila*), similar to the French research that produced the new Biscarrosse elm. One Italian focus is on finding cultivars that, in addition to having good disease resistance, are able to thrive in hot Mediterranean locations, and may be resilient to the extremes of weather coming as the climate changes.

In Spain there has been great reluctance to cross the native field elm with Asian species because there is concern that the Siberian elm, which has been in the country since the sixteenth century, is invasive already. It hybridises with native elms in the wild and there are fears that it will gradually colonise the country. So there, the scientific focus, as in Scotland, has been on identifying native trees that have survived the Dutch elm disease pandemic and breeding from them. They are also carrying out genetic analysis to try to establish exactly what is special about the DNA of elms that don't catch the blight.

All of these European efforts are combined through scientific collaboration at a continental level, through a European Union project which gives greater scope for cross-breeding among trees from different countries, in recognition that the fungus recognises no national borders.[9]

In America, elm breeding programmes at the United States Department of Agriculture are crossing American and European elms. At the University of Wisconsin, they are focusing on American natives and the Morton Arboretum in Chicago is exploring Asian options.

Without doubt there is still significant hope for elms in Canada, particularly with the huge area of the country – Alberta and British Columbia – where Dutch elm disease has yet to make any real impression. Meanwhile, Canada is also carrying out careful work to propagate trees that show resistance to the disease, giving evolution a little helpful shove and perhaps making amends for the negative interference that human society has caused. In Guelph, Ontario, a wealthy bird enthusiast with the beautifully appropriate name of Philip Gosling was so upset about the loss of the elms which were home to much-beloved visiting orioles that he offered half a million dollars to a scientist to try to help combat elm diseases.[10] At the campus of the University of Guelph, one elm tree appeared to have survived several bouts of Dutch elm disease unscathed, and professor of plant agriculture Praveen Saxena took up the challenge to get to the bottom of what made this resilient tree so special. He and several research students and post-docs have been doing genetic analysis of the tree and also cloning it.

So, all around the world there are substantial efforts to give elms a future after Dutch elm disease. There are other diseases, of course, to worry plant pathologists. One they're keeping their eyes on is a bacterium called elm yellows (*Candidatus Phytoplasma ulmi*), which is destructive to elms in North America, though so far less so in Europe.[11] The elm leaf beetle (*Xanthogaleruca luteola*) is native to Europe and not really a problem, primarily because it is preyed on and kept in check by the *Oomyzus gallerucae* species of wasp, but it is invasive and worrying in other parts of the world. No doubt there will be new invasive pests in future, the trees, insects and fungi will continue their co-evolutionary dance, and there will be increasing storms and droughts to contend with. That's life.

On 29 April 2024, I returned to Beauly. Gathered there was a small group of people: some officials from Historic Environment Scotland and NatureScot dressed in tweeds, looking the part of government officials; some botanists in sensible rain gear, wise given the forecast; plus a few artists and writers recognisable by their inappropriate footwear. There was the obligatory cameraman with a huge lens and commanding manner. Where the old elm tree had stood was a patch of bare ground, now to be occupied by one of its offspring, which had been spotted growing nearby and transplanted. It was perhaps three years old, knee-height, with stout-looking branches and a determined air. I wondered how it felt to be literally taking the place of its mother.

Then, more than 10 metres away, to prevent any risk of their roots linking up and potentially spreading disease in future, Max Coleman planted a very special first-year seedling, another of the 'potentially resilient' elms from the Royal Botanic Garden Edinburgh. Photographs were taken. There were no grand speeches or performances, but Historic Environment Scotland had to give themselves special permission to plant these youngster trees, as they have a default ban on any planting close to protected monuments, particularly as the ground is full of graves. After Max had moved away, Emma, who had grown the tree, came across and stamped it in more firmly. We took turns to fall on our knees with our cameras to take the little one's photograph and wish it well in the world. It's hard to imagine this slip of a thing, barely 15 centimetres tall, with its first few leaf buds opening gingerly into the world for the first time, potentially living for hundreds of years and growing to have a girth of many metres, but so it may.

This team is particularly interested in genetic analysis, and as I listened to them discuss genetic markers and genome mapping, I was soon out of my depth. They vary in their levels of optimism that this research will enable them to find out what makes some trees disease-free. The holy grail is to identify genes for Dutch elm disease resistance. There is no easy answer to the question and it looks like perhaps many different genes play a combined role in enabling a tree to defend itself by resisting fungal infection, or by evading it by tasting bad to beetles, for example.

The photographer corralled everyone under what is these days the biggest tree in the priory grounds, an imposing sycamore, with several nests in its canopy and one particularly irate blackbird who was not at all impressed to be playing host to a gathering of humans. Presumably there were chicks who needed protection. We were lined up in two rows with some jostling, and while we shuffled we soon noticed that a little above head height was a genuine wonder. Growing in a cleft between branches of the sycamore was a small but mature elm tree, complete with an impressive complement of samaras. Everyone was equally amazed, as it is highly unusual for elm to be epiphytic, growing on the surface of another plant, and it added yet another layer of delight to an already cheering event.

Good things come in threes, don't they? Now Beauly Priory has three young elms – one a child or grandchild of the Guardian of the Gateway; one a much-studied, special, 'potentially resilient' seedling; and one, wind-sown presumably, hiding up a tree.

I wish them well and have written them a little poem.

> Will these leaves, tender stems
> and delving roots
> have the power to recover our woods,
> to whelm our land
> with blight-surviving lushness?

Towards the end of May, after a couple of stretches of gorgeous weather, with temperatures soaring worryingly to high enough for elm bark beetles to fly, the elm seeds were ripening on the trees in Assynt. I had asked Nick Clooney, manager of the community tree nursery, to alert me when they were ready for collection. He'd told me months back that it's crucial to be watchful because they are so hasty in their maturing process. After growing gently over two months, the green, fleshy, unripe clusters suddenly dry out and fall within a matter of days. After bursts of sunny, dry, breezy weather, we had a day and night with plenty of rain on the back of a full gale, and the following morning a text message pinged in from Nick, letting me know that he and his colleague Josie were heading off for elm seeds. In Lochinver, close to the seashore, the seeds were already well scattered, the wind having torn them from their twigs, but further inland on slightly higher ground, the trees were clearly a few days behind. On some trees the bunches of samaras were still green and fruity-looking, and Nick declared that they would wait for another week. Some locals would gather these seeds at a community open day, encouraging more people to become interested in the future of elms in our landscape.

Further up the glen, with the Traligill River cascading by at the bottom of its treacherously steep gorge, there were trees with plenty of papery samaras rustling in the breeze, which was much reduced from the gale the night before but just sufficient to keep the midges at bay. Both Nick and Josie had been expecting to be under attack by our famous biting insects and were dressed in full protective gear, like beekeepers in camouflage, and both Bill and I had our midge hats in our pockets. But we were in luck and the light breeze meant we were free to stroll unmolested from tree to tree up above the ravine, tugging off handfuls of the fruits from the twigs on which they grow so prolifically and stuffing them into Nick and Josie's carrier bags, stopping frequently to point out to each other beetles, shield bugs, a lacewing that Nick hadn't seen before, the water avens coming into flower, the masses of early purple orchids in full glory and a badger's latrine. Nick's dog padded around amiably.

The elm fruits are like miniature spaceships, each samara formed of a beige papery outer circle around the central seed, a brown or in some cases distinctly pink bump, from which a whole new tree may sprout. And like a spaceship, they managed to transport me. Before I'd set out I'd been feeling low, but by lunchtime the elm trees had performed their magic and I felt restored to hopefulness.

Nick and Josie have been growing elm trees at the nursery for a decade. Every year they sell elms to local tree planting and woodland establishment projects, who put them in the ground. We have hundreds of hectares of new native plantations in the parish and most of them have elms growing in them. Each of these gives hope for the future of the species in our area.

Even if we do get a wave of Dutch elm disease and it kills some of our giant trees in the next few years, most of these planted trees will be too young to host the elm bark beetles and thus be affected. Even if the disease persists in the area, some of these thousands of trees, all of which are slightly genetically different, will survive. If we lose some big old elms, they will slow down a river and help the fish, or rot away and support some interesting fungi. They'll be home to insects that will feed birds and badgers, and so their positive role in our woodland ecosystem will continue. Death isn't a bad thing. Life carries on. As individuals they will not die in vain, and as a species in our landscape, I feel confident that elm will find a way to thrive.

It was a balmy day for a woodland open day in early June, when four more 'potentially resilient' elms were planted out from the Royal Botanic Garden Edinburgh, along with a slideshow from Max Coleman and a reading from a children's book about elms by Euan Bowditch. Nick led a walk up the glen and we gathered seeds for the nursery. Children ran and played in the woods and learned how to recognise an elm tree when they bumped into it. It was a joyful day and easy to feel optimistic.

We took the opportunity to have the visiting experts examine the trees we had been worrying about. Max sent me a message later.

'I took a drive down there after the event and can see why there was concern. At least two trees are showing dieback and a stag-headed look. However, the good news is that this is not due to DED. These are mature elms and they have shown relatively minor dieback. Old trees will do this for all sorts of reasons, so it's hard to say why in this case. The key thing is that there are no symptoms suggestive of DED.'

This year, the elms that I was worried about last year are not any worse. If it was Dutch elm disease, they would be declining fast, wilting and withering, but they're not. I'm convinced now that some of the yellowing was just early autumn colour, and that the leafless limbs are simply old.

We believe that pruning of some branches off elms that were shading an office at Glencanisp may have been the cause of a localised infection. Perhaps the chainsaw was carrying spores of *Ophiostoma nova-ulmi* or some came in on the feet of passing walkers. The five affected elms are all close enough together for the fungus to spread through their gregarious roots connected underground. Although it's dreadful to lose those trees, and to have the fungus present in the area, if it arrived by human transmission rather than by beetles, there is a good chance that it may remain an isolated outbreak. For now at least, the rest of our elms in the woods and up the glens are flourishing. It feels like the winter of 2020, the first year of the COVID-19 pandemic, when the virus had yet to reach our community, until a couple of people who had been working away arrived home with the infection just before Christmas. They quarantined, and everyone waited to see who would let it loose into the local population. But their quarantine was tight, their respect for everyone else strong; they recovered and managed not to infect anyone else, and we all carried on, a bubble of 'Novid' people. Eventually a more virulent form swept through and took most of us down, but until then we felt blessed, escapees in a refuge from a trouble afflicting everyone else.

And here we are again, only this time we're a refuge from Dutch elm disease, and once again we're not complacent. We're using the time we have to increase the resilience of our woods and to offer an optimistic alternative to the mainstream story of doom. Yes, across the world, we've lost a lot of elms, but we haven't lost hope.

Elmo

One of the best-known and loved trees in our area is a huge elm on what's fondly known as The Higgledy Piggledy Path, one of a network of walking routes through our community woodland, the Culag Woods. This small area of mostly conifer plantation, close to the harbour in Lochinver, belongs to the wealthy Vestey family who own much of our parish, including key buildings in the village. Back in the 1990s, after storms blew down many of the trees, blocking popular dog-walking paths, a group of local people took over the woods to reverse the neglect that it had been subjected to for many years. The Culag Community Woodland Trust negotiated a peppercorn rent of the land for ninety-nine years and set about clearing and rebuilding paths, creating new routes through the woods, making a play area, generating firewood from wind-thrown trees, organising social events, running training courses and generally turning the area into the thriving community asset it is today.

The Higgledy-Piggledy Path runs from roughly the centre of the woods, meandering down and up and around and about and up and down to the harbour. Not far from its start, right in the heart of the woods, is an elm tree. It is the tallest one I know, reaching out above the canopy of the conifers, among a pocket of broadleaf trees left on a steep, bouldery crag deep inside the plantation. It and the conifers shelter each other. There is another, older and more fragile-looking, multi-stemmed elm beside it, some smaller, younger or hungrier ones further up the slope, and one that might be really ancient, clinging to rock in that desperado pose elms do so well. Among them are birches, hazels, an ash and an oak, a holly, some rowans and maybe other species too. But I've come to see the big elm. The forecast was for a bright day but so far it is cold and overcast.

It's early June and the bole is a fuzz of green shoots, deer-browsed and small-leaved. Mosses harbour the trefoil leaves of wood sorrel. Bluebells, finished flowering now, flop around the roots where sedges are taking their place among lady ferns and hard ferns and some shade-enfeebled grasses. Nothing is flowering under this elm at all. This is the time of year when bees starve, after the mass spring blossoming of rowans and fruit trees but before the brambles and heather that will feed them through the later summer. There are a couple of foxgloves and some gleaming yellow spearwort just outside the canopy, lit by a brief burst of sunshine. It might be warmer there but I don't want to leave the tree.

Chris Puddephatt sent me a photo of this elm a few days ago with the message 'lush, lush, lush!' and it reminded me to gather some of its seeds. In the depths of winter, I sat here with Mairi, one of the people who helped me when I felt almost out of my mind trying to deal with horrors from my childhood that had remained buried until my father was dead. I was feeling completely broken, and she suggested a walk, so we came here and I showed her the tree and talked about the ferns and *Lobaria* lichens that festoon it so dramatically through the months when it lacks its own foliage. We saw how the tree is a host to all these other life forms which photosynthesise through the months when the tree cannot, soaking up the rain and fixing carbon and nitrogen from the air and gradually raining it down onto the soil below to benefit everything else in the ecosystem. This tree, like all elms, is not just a single, independent entity. It is part of a living community. The mosses, ferns and lichens are not 'parasites' on it, nor just passively hitching a ride through life. They are all mutually supportive. Somehow that conversation supported me too, and helped to confirm that I could still function, despite struggling. Returning here, I see that some of my most difficult feelings have shifted in the intervening time. I am healing gradually, and elms have been helpful medicine.

I came here with a group of primary school children and we measured the canopy, which covers an area of more than

175 square metres within which we discovered dozens of mosses, liverworts, ferns and lichens. They have befriended the big furry tree and nicknamed it Elmo!

Today, the elm is lush indeed, with twigs of leaves sprouting effusively from its branches, a luxuriance of green-gold foliage tessellating to almost entirely blot out the grey, rain-threatening sky. These leaves are mopping up all of the feeble light our cold, wet June can offer. The branches are still dressed in lichens and ferns, frilly underwear beneath the green gown, but they are paler on their undersides and more desiccated than they were in winter. Though damp, this rainforest is not quite so wet at this time of year, and the leaf-cover acts as an umbrella, so the epiphytes shrink back – it's not their season.

The wind is hushing and shushing, whispering and chuckling up in the canopy. A chaffinch punctuates the breeze-speech with a repeated long trill ending with a question, long trill then question: It's-like-this, don't-you know? More distant birds twit briefer responses. I listen in to the chatter, an uncomprehending audience, as if in a café in a foreign country, just enjoying being there, vaguely getting the gist of what it's like here and if I don't like it I can go home. I do like it, but I'll go home soon anyway. I like it there too.

Before I go, I breathe in the scent of the elm. It smells deep and clean. Lichens on the bark are dusty-white. Seeds are hanging from spider gossamer threads, suspended in webs and scattered everywhere on the ground. I run my fingers down a crack in the bark on the trunk, which is ridged like corduroy for a giant, though the underside of the branches is smooth. All possible textures are here to be explored. I see that the tree has suffered much loss. Branches fallen have left scars. The business of being home to myriad insects and birds, not to mention standing there in all weathers, has left it haggard-skinned, rough and scarred, but still very much alive. I reach for that sense of life, that strength burgeoning up from its roots, its ongoing growth despite wild weather and the intrusions of others. It lives and grows because of it all – with, through and beyond it all. It bears difficulty and flourishes anyway. I tell the tree to keep cool and hope that this shady environment will remain inhospitable to elm bark beetles for years to come. After one last touch, I thank the tree and walk back towards the harbour. At long last, the sun is coming out.

Appendix 1

'Elm tree talk'[1]

I am *leamhan*, wych elm, *Ulmus glabra*,
orme, *iep*, rüster, вяз (vyas), *jalava*.[2]
My life spans many of yours.[3] I see
deep into the earth and high into the sky.
I do not move as you do, but please
don't mistake my standing stillness
as any kind of lack. I spread underground.[4]
I dance with gales and storms,
scatter my leaves[5] and twigs and seeds.
Sometimes I see humans value

---

1. I was commissioned by Circus Arts to write a response to the elm tree in Beauly Priory grounds for the ceremony 'Guardian of the Gateway' on 24 September 2022. I expected to write an essay of some sort but instead found myself giving voice to the tree itself. I hope that it doesn't mind. The 'rootnotes' are from my own point of view.

2. Scottish Gaelic, English, Latin (scientific), French, Dutch, German, Russian, Finnish.

3. How much trees must see!

4. Indeed this is, for some elms, their usual means of self-propagation.

5. 'When elm leaves are as big as a shilling,
   plant your beans if to plant them you're willing.
   When elm leaves are as big as a penny,
   plant your beans if you mean to have any.' (Anon)

movement over growth,
growth over rhythm,
but rhythm is everything.
I've been rocking to the pulse of sun and moon
for centuries.[6] Listen:

Tree time is not like your time.
I have watched you for 800 sun-cycles,[7]
concluded long ago that you are strange
creatures of movement and desire.
Still yourselves for a moment, please.
Put down all your wishes and longings,
all your cravings and dissatisfaction.
Lighten yourself and breathe our gift:
oxygen,[8] the one thing you can't do without
but often seem to leave off your wishlist.
But I'm not here to criticise, I don't want . . .

---

6. Natural time is cyclical: tides and days and seasons go round and round. The nave (centre) of the oldest wheel found in Scotland, more than 2000 years old, was made of elm.

7. Isn't it exciting that we have historical records to show that the tree is this old? Planted in the early days of the priory in one of two rows of elms that used to grace the pathway towards the now ruined building, this tree has long outlived the human edifice for which it was the gateway. It is managed by Historic Environment Scotland and let's hope it continues to be looked after and allowed to die with dignity.

8. Trees, like all plants, thanks to the miraculous marvel of photosynthesis, breathe in carbon dioxide through little mouths in their leaves, called stomata, and breathe out oxygen. This is the reverse of what we do. If we're ever feeling disconnected to nature, we just need to breathe in and there we are – full of tree breath!

well, I don't want anything,[9] least of all
you feeling bad or not wanting to listen.
So. Breathe. Settle. Tune into tree talk.

I wish for nothing, I do not know desire,
no dreams or dreads,
no hopes, no disappointments.
I stand and the world is here
in all its diverse, subtle
and endlessly changing beauty.[10]
I wish for nothing because I am full,
from crown to root tip,[11] with wonder.

I say I, because you expect a voice
of singular identity, which seems to be
how you perceive yourselves.[12]
But I am really we.
I am, we are, a community.

---

9. As this ancient tree becomes ever more frail, it speaks to me of the fragility of dementia, a strange blessing of which can be a lessening of the ability to cling onto desires. In the loss or change of the self that dementia brings, there comes, sometimes, an ability to live purely in the present moment. I speculate that a tree probably has that ability from the word go.

10. Sight is the first sense we can use to appreciate the tree. Its form is skeletal now, its branches mere sticks on the barrel of its vast trunk. Yet still it commands its place, filling our view, looking magnificent. Half of it, perhaps more than half, all of its underground depth, is hidden. Even with our strongest sense we can only perceive a part of its totality.

11. Elm leaves heal wounds; elm roots heal bones.

12. Elm twigs are supposed to help to produce eloquence, and prevent idle gossip.

Though you may see me as just a tree
I am wasps and willowherbs,
spiders and their silken threads,[13]
netted midgies, mosses, lichens,
an asylum-seeking cherry tree[14]
brought by one of my many birds,
a bounty of boring beetles[15]

and in the soil a riot of life
I can't begin to tell you of
for if I do I'll enter a fungal epic[16]
that will keep you here
until your leaves fall for the last time too.
It is enough to say they smell divine![17]

---

13. A passing cat appeared briefly in the poem, settled for a while in these rootnotes, then moved on . . .

14. It may be a cotoneaster, though my contact at NatureScot couldn't confirm one way or the other. Until a botanist gets a ladder up there to find out, we won't know exactly what's growing on the tree. It's good to be uncertain. A bit of mystery is a fine thing.

15. Dutch elm disease, the fungal infection (*Ophiostoma novo-ulmi*) that is killing the tree, has been conveyed to it by elm bark beetles (*Scolytus scolytus*), which are slowly making their way north as the climate warms.

16. There is no way of overestimating the importance of fungi in this 'entangled life' that we all live, as Merlin Sheldrake so brilliantly explains in his book of this name. Elm is a symbol of death, the tree of elves, linked with Orpheus who set off down into the underworld to bring back Eurydice, so it's an ambiguous kind of death that is very much alive. Fungi know all about this.

17. The second of our senses with which we can appreciate the tree is scent. The tree's bark, on a warm afternoon, smells spicy, reminiscent of cinnamon.

Breathe in their petrichor
after a downpour of blessed rain,
I swear you'll never feel the same.
Good old-fashioned air,[18]
it's better than prayer for getting you
to a state of peace or even bliss.

I've often felt bliss –
it is the sun's best present,
that feeling of being drenched in sweetness.[19]
Did I mention the birds?
All those songs we've sung!
That's one thing your monks[20] did really well,
music billowing from that building[21] over there,
booming and brilliant
or murmuring yet still melodious,
heartfelt harmonies and tunes that filled
my phloem[22] and still flow within me.

---

18. Breathe! There we are. Connected again.

19. Taste is the next sense with which we can appreciate the tree. Traditionally, the inner bark of elm is used to create a soothing milk substitute, liquid sunshine, good for guts.

20. Beauly, in Gaelic, is *A' Mhanachainn*, literally The Monks, after the Valliscaulian Order who made it their home, and who planted the elm.

21. The priory, built by the monks in 1230.

22. Trees are full of tubes. The phloem are those through which the sweet food made from the leaves flows to the rest of the plant. They also have xylem, which are tubes through which water, sucked up from the ground, flows upwards to irrigate all their thirsty bits.

Keep on with your music,[23] people,
your crafts of making our wood into pipes,
or fiddle and bow, cello or oboe –
don't ever let that go! Please, you bring your gifts
into the world and everyone on Earth
knows that's what you do.
Paintings, sculptures, knitted wool,
boats,[24] those filigrees of silver, all your arts are fine,
and for that we forgive you everything,
and we all know there's plenty needs forgiving.
The CO_2 – I've done my best to, as you might say,
'suck it up', but steady on.[25] That's all I'll say
because I'm here to praise, to thank the earth
for giving me this life, these lives, our lives.
I'll not go on much longer.
I just want to let you know I feel for you
fumbling your ways into the future.[26]
Yes, I feel for you and feeling
is where we're just the same.

---

23. Sound is a great sense for appreciating trees. They are in constant dialogue with wind, of course, but it is always worth putting your ear to a trunk and listening to some quiet wisdom.

24. In boat building, elm is used for the garboard strake, the first wale – not to be confused with the whales, who are down below – laid next to the keel.

25. Who knows what our excess emissions will cause? Isn't the demise of this tree enough to make us pause?

26. Dismantling capitalism may be the only way to transition to net zero carbon emissions. Wish us luck, please, tree!

Touch[27] my sun-warmed wood
with your skin and I touch you.
Stroke me, hug me,
and our pleasure is mutual.
We become one.

---

27. Our final sense for appreciating the tree is the best one: touch. The tree has so many textures: dimpled, creviced and fissured, fluffy with webs, dusty from the work of the beetles and wasps inside, smooth as a well-turned doorknob. You can use your whole body to feel its strength and share in solidarity.

Appendix 2

The Elm Family

This is a list of elm species organised according to their genera, subgenera and sections, but note that there is ongoing dispute about the sections within the *Ulmus* subgenera. The list draws primarily from Fragnière et al. (2021).[1] Other sources include Whittemore et al. (2021).[2] Some common or alternative names for *Ulmus* species are included. These sometimes do not apply to the whole species, being used for particular varieties (e.g. English elm only applies to *Ulmus minor* 'Atinia') or only to some geographical locations (e.g. white elm is usually only used for *Ulmus americana* in Canada).

Genus Ampelocera
- *Ampelocera albertiae*
- *Ampelocera cubensis*
- *Ampelocera edentula*
- *Ampelocera glabra*
- *Ampelocera hottlei*
- *Ampelocera longissima*
- *Ampelocera macphersonii*
- *Ampelocera macrocarpa*
- *Ampelocera ruizii*

Genus Phyllostylon
Phyllostylon brasiliense
Phyllostylon rhamnoides

Genus Holoptelea
Holoptelea grandis
Holoptelea integrifolia

Genus Hemiptelea
Hemiptelea davidii

Genus Planera
Planera aquatica

Genus Zelkova
Zelkova abelicea
Zelkova carpinifolia
Zelkova sicula
Zelkova schneideriana
Zelkova serrata (keyaki)
Zelkova sinica

Genus Ulmus
Subgenus Indoptelea
Ulmus villosa – Marn elm, cherry-bark elm
Subgenus Oreoptelea
  Section Blepharocarpus
Ulmus americana – American elm, white elm
Ulmus laevis – European white elm, fluttering elm, spreading elm, Russian elm

Section Chaetoptelea
- *Ulmus alata* – winged elm, wahoo
- *Ulmus crassifolia* – cedar elm
- *Ulmus elongata* – long raceme elm
- *Ulmus ismaelis* – Ismael's elm
- *Ulmus mexicana* – Mexican elm
- *Ulmus serotina* – September elm
- *Ulmus thomasii* – rock elm, cork elm

Subgenus Ulmus

Section Foliaceae
- *Ulmus castaneifolia* – chestnut-leafed elm, multinerved elm
- *Ulmus changii* – Hangzhou elm
- *Ulmus chenmoui* – Chenmou elm, Langya Mountain elm
- *Ulmus chumlia*
- *Ulmus davidiana* – David Elm, Father David's elm, Japanese elm
- *Ulmus harbinensis* – Harbin elm
- *Ulmus microcarpa* – Tibetan elm
- *Ulmus minor* – field elm, grey elm, grey-leafed elm, hoary elm, Mediterranean elm, English elm
- *Ulmus prunifolia* – cherry-leafed elm
- *Ulmus pumila* – Siberian elm
- *Ulmus szechuanica* – Szechuan elm

Section Microptelea
- *Ulmus lanceifolia* – Vietnam elm
- *Ulmus parvifolia* – Chinese elm, lacebark elm
- *Ulmus mianzhuensis* – Mianzhu elm

Section Trichocarpus
- *Ulmus glaucescens* – Gansu elm
- *Ulmus lamellosa* – Hebei elm
- *Ulmus macrocarpa* – large-fruited elm

Section Ulmus
- *Ulmus bergmanniana* – Bergmann's elm
- *Ulmus glabra* – wych elm, Scots elm
- *Ulmus laciniata* – Manchurian elm, cut-leaf elm, Nikko elm
- *Ulmus rubra* – slippery elm, red elm
- *Ulmus uyematsui* – Alishan elm
- *Ulmus wallichiana* – Himalayan elm, Kashmir elm

Section Incertae sedis
- *Ulmus gaussenii* – Anhui elm
- *Ulmus pseudopropinqua* – Harbin spring elm

Suggested Further Reading

Clouston, B. & Stansfield, K. (Eds.). (1979). *After the Elm . . . .* William Heinemann Ltd.

Coleman, M. (2009). *Wych Elm.* Royal Botanic Garden Edinburgh.

King, A. & Clifford, S. (Eds.). (1989). *Trees Be Company.* Green Books.

Richens, R. H. (1983). *Elm.* Cambridge University Press.

Seddon, M. & Shreeve, D. (2024). *Great British Elms.* Kew Publishing.

Somerville, R. (2021). *Barn Club.* Chelsea Green Publishing Company.

Sticklen, M. B. & Sherald, J. L. (Eds.). (1933). *Dutch Elm Disease Research.* Springer-Verlag.

Wilkinson, G. (1978). *Epitaph for the Elm.* Arrow Books.

Acknowledgements

This book would never have happened but for Kirsten Boddy and Isabel McLeish, who asked me to write about the dying elm in Beauly in 2022 – thank you both. The subsequent odyssey around the world of elms has involved encounters with so many enthusiastic and helpful people, and I am especially grateful to Max Coleman and his colleagues at the Royal Botanic Garden Edinburgh, Euan Bowditch and colleagues at the Scottish School of Forestry, Alister Peter, Rachel Blow, Gareth Parkinson and Sir Martin Holdgate for specialist knowledge. Locally, Chris Puddephatt has been an endless inspiration with his magnificent photographs, and I'm grateful also to Ian Evans, Nick Clooney, Josie Gibberd, Andy Summers, Kat Martin, Alison Roe, Jane Matheson, Romany Garnett, Elaine and Lewis MacAskill, Richard Williams, Jorine van Delft, Fiona Saywell, Gwen Richards, Stuart Belshaw, Fritha West, Lynn Bennett-Mackenzie, Anna Mackay, Chris Goodman, Rehema White, Jack Travers, Althea Davis and everyone else who has contributed to the Assynt Elm Project.

Thanks to the team at Wildfire, especially my editor Philip Connor, for being so tirelessly keen, kind and professional throughout the process, Lindsay Davies for commissioning the book and DeAndra Lupu for her brilliant editorial advice and attention to detail. Thanks also to Cailtin Raynor, Jo Edwards,

Federica Leonardis and all the others doing the magic behind the scenes to make this book possible.

Fiona Hamilton, Mairi Mackay, Hazel Pepper, Maria Hynes and Tracy Kennedy, I can't thank you enough for your support through difficult times. Most of all, love and thanks for endless kindness, tree-hugging and laughter, to my husband, Bill Ritchie.

References

One – Death: Introduction

1 Karnosky, D. F. (1979). Dutch elm disease: A review of the history, environmental implications, control and research needs. *Environmental Conservation*, 6(4), 311–22.
2 Hutchison, P. J. (2012). When Elm Street became treeless: Journalistic coverage of Dutch elm disease, 1930–80. *Journalism History*, 38(2), 100–9.
3 Heybroek, H. M. (1993). Why bother about the elm? In M. B. Sticklen & J. L. Sherald (Eds.), *Dutch Elm Disease Research: Cellular and Molecular Approaches*. Springer-Verlag.
4 American Elm Heritage Project information series. (2011, January 6). Forks Forum. www.forksforum.com/news/american-elm-heritage-project-information-series
5 History Cambridge. *Washington Elm debate rages on: Fact or legend?*. historycambridge.org/articles/washington-elm-debate-rages-on-fact-or-legend

Two – Life: Ecology

1 Fragnière, Y., Song, Y.-G., Fazan, L., Manchester, S. R., Garfi, G. & Kozlowski, G. (2021). Biogeographic overview of Ulmaceae: Diversity, distribution, ecological preferences, and conservation status. *Plants*, 10(6), 1111.
2 Coleman, M. (2024, March 26). *Elm blossom*. Botanics Stories. stories.rbge.org.uk/archives/38649
3 Edlin, H. L. (1958). *The Living Forest*. Thames and Hudson.
4 Wageningen University & Research. *Image collections*. images.wur.nl/digital/collection/coll13/search/searchterm/Ulmus

5 Coleman, M. (2009). *Wych Elm*. Royal Botanic Garden Edinburgh.
6 Marks, C. (2016). The ecological role of American elm (*Ulmus americana* L.) in floodplain forests of northeastern North America. Conference: Proceedings of the American Elm Restoration Workshop. www.fs.usda.gov/nrs/pubs/gtr/gtr-nrs-p-174papers/10marks-gtr-p-174.pdf
7 Fuller, R. E. (2017, November 25). How an elm tree that died decades ago is still giving life today. Robert E Fuller Wildlife Artist. www.robertefuller.com/blogs/blog/how-an-elm-tree-that-died-decades-ago-is-still-giving-life-today
8 BirdLife International. *Red-headed woodpecker*. https://datazone.birdlife.org/species/factsheet/red-headed-woodpecker-melanerpes-erythrocephalus

Three – Death: Dutch Elm Disease

1 Edlin, H. L. (1958). *The Living Forest*. Thames and Hudson.
2 Karnosky, D. F. (1979). Dutch elm disease: A review of the history, environmental implications, control and research needs. *Environmental Conservation*, 6(4), 311–22.
3 Grylls, T. & van Reeuwijk, M. (2022). How trees affect urban air quality: It depends on the source. *Atmospheric Environment, 290*.
4 Karnosky, D. F. (1979). Dutch elm disease: A review of the history, environmental implications, control and research needs. *Environmental Conservation*, 6(4), 311–22.
5 Richards, W. C. (1993). Cerato-ulmin: A unique wilt toxin. In M. B. Sticklen & J. L. Sherald (Eds.), *Dutch Elm Disease Research: Cellular and Molecular Approaches*. Springer-Verlag.
6 Osborne, P. (1985). Some effects of Dutch elm disease on the birds of a Dorset dairy farm. *Journal of Applied Ecology, 22*(3), 681–91.
7 Carpintero, S. & Reyes-López, J. (2020). Indirect effect of the invasive exotic fungus *Ophiostoma novo-ulmi* (Dutch elm disease) on ants. *Community Ecology, 21*(2), 133–43.
8 Peglar, S. M. & Birks H. J. B. (1993). The mid-Holocene *Ulmus* fall at Diss Mere, South-East England – disease and human impact? *Vegetation History and Archaeobotany, 2*, 61–8.
9 Flynn, L. E. & Mitchell, F. J. G. (2019). Comparison of a recent elm

decline with the mid-Holocene Elm Decline. *Vegetation History and Archaeobotany*, 28(4), 391–8.
10 Clark, S. H. E. & Edwards, K. J. (2004). Elm bark beetle in Holocene peat deposits and the northwest European elm decline. *Journal of Quaternary Science*, 19, 525–8.
11 Artists Open Houses. *Elpida Hadzi-Vasileva and the Preston Park Twins*. aoh.org.uk/elpida-hadzi-vasileva-and-the-preston-park-twins

Four – Life: Healing Uses

1 University of Minnesota Extension. *Yard and garden*. apps.extension. umn.edu/garden
2 iNaturalist. *Elm balloon-gall aphid (Eriosoma lanuginosum)*. www. inaturalist.org/taxa/416160-Eriosoma-lanuginosum
3 Johnston, G. (1853). *The Botany of the Eastern Borders*. John Van Voorst.
4 Medicine Traditions. *Ulmus, elm tree*. https://www.medicinetraditions. com/ulmus-elm.html
5 Hopman, E. E. (1991). *Tree Medicine, Tree Magic*. Phoenix Publishing.
6 Bach, E. (1941). *The Twelve Healers and Other Remedies*. The CW Daniel Company.
7 Ibid.
8 Martynoga, F. (2012). *A Handbook of Scotland's Wild Harvests*. Saraband.
9 Jie, X. (2024). Xia Jie from Shanbei. https://www.youtube.com/channel/UCV-s8tichDSSc19lUw08kXQ
10 A Food Forest in Your Garden. www.foodforest.garden
11 Bowditch, E. A. D. & Macdonald, E. (2021). The elm is dead! Long live the elm! New developments for elm conservation in Scotland. *Scottish Forestry*, 75(2), 29–38.
12 Coleman, M., A'Hara S. W., Tomlinson, P. R. & Davey, P. J. (2016). Elm clone identification and the conundrum of the slow spread of Dutch elm disease on the Isle of Man. *New Journal of Botany*, 6(2–3), 79–89.

Five – Death: Coffins and Cartwheels

1 Edlin, H. L. (1958). *The Living Forest*. Thames and Hudson.
2 Richens, R. H. (1983). *Elm*. Cambridge University Press.
3 Evans, G. E. (1956). *Ask the Fellows Who Cut the Hay*. Faber & Faber.
4 Somerville, R. (2021). *Barn Club*. Chelsea Green Publishing Company.

5 Tudge, C. (2005). *The Secret Life of Trees*. Penguin.
6 Stafford, F. (2016). *The Long, Long Life of Trees*. Yale University Press.
7 Alrune, F. (1992). A Mesolithic elm bow approximately 9000 years old. *Journal of the Society of Archer-Antiquaries, 35*, 47–50.
8 Stansfield, K. (1979). The elm in the service of man. In B. Clouston & K. Stansfield (Eds.), *After the Elm* . . . William Heinemann Ltd.
9 Woodland Trust. *Elm, Huntingdon*. https://www.woodlandtrust.org.uk/trees-woods-and-wildlife/british-trees/a-z-of-british-trees/huntingdon-elm/
10 Mackenzie, D. (1996, December 7). Early Dutch dam builders were plagued by lice. *New Scientist*. https://www.newscientist.com/article/mg15220590-400-early-dutch-dam-builders-were-plagued-by-lice/
11 Milliken, W. & Bridgewater, S. (2004). *Flora Celtica*. Birlinn Ltd.
12 Milman, O. (2022). *The Insect Crisis*. Atlantic Books.
13 All material for this vignette is sourced from Coleman, M. (2009). *Wych Elm*. Royal Botanic Garden Edinburgh.

Six – Life: Elms Around The World

1 Darwin, T. (2008). *The Scots Herbal*. Birlinn Ltd.
2 UNESCO. *Chinese paper-cut*. ich.unesco.org/en/RL/chinese-paper-cut-00219
3 Institute of Applied Food Allergy. *Chirabilva (Holoptelea integrifolia)*. https://www.iafaforallergy.com/herbs-a-to-z/chirabilva-holoptelea-integrifolia/
4 Mondal, S. & Bandyopadhyay, A. (2016). The wonders of a medicinal tree: *Holoptelea integrifolia*. *International Journal of Pharmacy and Pharmaceutical Sciences, 8*(8), 43–8.
5 De Roerich, G. (1931). *Trails to Inmost Asia*. Yale University Press.
6 Hiroshi, T. (2017, December 8). Slumbering giants: Three old-growth trees in early winter. https://www.nippon.com/en/views/b05307/
7 Heian Period Japan. *Keyaki zelkova tree legends*. heianperiodjapan.blogspot.com/2016/08/keyaki-zelkova-tree-legends.html
8 Wilkinson, G. (1978). *Epitaph for the Elm*. Arrow Books.
9 Clouston, B. & Stansfield, K. (Eds.). (1979). *After the Elm* . . . William Heinemann Ltd.
10 Danin, A. & Fragman-Sapir, O. *Ulmus minor Mill*. Flora of Israel Online. flora.org.il/en/plants/ULMMIN

11 Thompson, K. (2014). *Where do Camels Belong?* Profile Books.
12 Ellis, M. Grange Farm. www.arboreta.nl/grangefarm.htm
13 Plantinfo. *Ulmus parvifolia (Chinese elm)*. plantinfo.co.za/plant/ulmus-parvifolia
14 The Elms. *History*. theelms.org.nz/history
15 Ganley, R. J. & Bulman, L. S. (2016). Dutch elm disease in New Zealand: Impacts from eradication and management programmes. *Plant Pathology*, 65(7), 1047–55.
16 Gadgil, P. D., Bulman, L. S., Dick, M. A., Bain, J. & Dunn, C. P. (2000). Dutch elm disease in New Zealand. In C. P. Dunn (Ed.), *The Elms*. Kluwer Academic Publishers.
17 Fatal Dutch elm disease found on Waikato property. (2024, March 5). RNZ. www.rnz.co.nz/news/national/510896/fatal-dutch-elm-disease-found-on-waikato-property

Seven – Death: European Elms and their Folklore

1 Gil, L., Fuentes-Utrilla, P., Soto, Á., Cervera, T. M. & Collada, C. (2004). English elm is a 2,000-year-old Roman clone. *Nature*, 431, 1053.
2 Tudge, C. (2005). *The Secret Life of Trees*. Penguin.
3 Richens, R. H. (1983). *Elm*. Cambridge University Press.
4 Ian. (2020, May 19). Maud's elm. Mysterious Britain & Ireland. www.mysteriousbritain.co.uk/legends/mauds-elm
5 Homer. (1898). *The Iliad* (S. Butler, Trans.). Longmans.
6 Ibid.
7 Protesilaus (2022). *Ancient Literature*. ancient-literature.com/protesilaus/
8 Virgil. (1956). *The Aeneid* (W. F. J. Knight, Trans.). Penguin.
9 Catullus, C. V. (2007). *The Carmina of Caius Valerius Catullus*. Project Gutenberg. www.gutenberg.org/files/20732/20732-h/20732-h.htm
10 Cato. (1998). *On Farming (De Agricultura)*. Prospect Books.
11 Heybroek, H. M. (2014). The elm, tree of milk and wine. *iForest – Biogeosciences and Forestry*, 8(2), 181–6.
12 Fuentes-Utrilla, P., López-Rodriguez, R. A. & Gil, L. (2004). The historical relationship of elms and vines. *Forest Systems*, 13(1), 7–15.
13 Ovid. (1955). *Metamorphoses* (M. Innes, Trans.). Penguin.
14 Ibid.
15 Beith, M. (2000). *A' Chraobh (The Tree)*. Dornoch Studio.

16 Newton, M. (2019). *Warriors of the Word*. Birlinn Ltd
17 Ibid.
18 Ainmean-Àite na h-Alba (Gaelic Place-Names of Scotland). *The database*. www.ainmean-aite.scot
19 Carmichael, A. (1992). *Carmina Gaedelica*. Floris Books.
20 Newton, M. (2019). *Warriors of the Word*. Birlinn Ltd.
21 Price, N. (2022). *The Children of Ash & Elm*. Penguin.
22 Heybroek, H. M. (2014). The elm, tree of milk and wine. *iForest – Biogeosciences and Forestry*, 8(2), 181–6.
23 Geusa, A. (2021, March 1). Olga Kisseleva: Deciphering the talk of trees. Art Focus Now. https://artfocusnow.com/people/olga-kisseleva-deciphering-the-talk-of-trees/

Eight – Life: American Elms

1 USDA. *Rock elm*. https://www.srs.fs.usda.gov/pubs/misc/ag_654/volume_2/ulmus/thomasii.htm
2 Texas Real Food. *The traditional uses of slippery elm in Native American medicine*. discover.texasrealfood.com/roots-of-wellness/slippery-elm
3 Biolifehacks. (2023, November 6). Slippery elm: The game-changing herb that's taking the wellness world by storm. *Medium*. medium.com/@biolifehacks/slippery-elm-the-game-changing-herb-thats-taking-the-wellness-world-by-storm-fe9f5b49f19b
4 Texas Real Food. *The traditional uses of slippery elm in Native American medicine*. discover.texasrealfood.com/roots-of-wellness/slippery-elm
5 Tree Pursuits. What Do Elm Trees Represent? (Exploring The Symbolism). https://treepursuits.com/what-do-elm-trees-represent/
6 Ciaravola, D. R. (2016, October 18). Oval trees 101: Do you know their history? Colorado State University. https://source.colostate.edu/oval-trees-101/
7 Monumental Trees. American elm. https://www.monumentaltrees.com/en/trees/ulmusamericana/records/
8 Duinker, P. (2020, June 1). The American Elm. Halifax Tree Project. www.halifaxtreeproject.com/the-american-elm
9 University of Prince Edward Island. *The Big Elm*. The Island Narratives Program. https://islandarchives.ca/islandora/object/cap%3A1256?solr_nav%5Bid%5D=a31e1de961f8ea069cd4&solr_nav%5Bpage%5D=0&solr_nav%5Boffset%5D=2

10. Manitoba Agriculture and Resource Development. (2020). *Dutch elm disease management in Manitoba.* https://www.gov.mb.ca/nrnd/forest/pubs/forest_lands/health/dutch-elm-disease-mgmt-2020.pdf
11. Trees Winnipeg. *Dutch elm disease.* www.treeswinnipeg.org/our-urban-forest/urban-forest-threats/dutch-elm-disease
12. Ministry of Agriculture, Food and Fisheries. (2021, March). *Dutch elm disease.* https://www2.gov.bc.ca/assets/gov/farming-natural-resources-and-industry/agriculture-and-seafood/animal-and-crops/plant-health/disease-factsheets/nursery-and-ornamentals/af_dutch_elm_disease.pdf
13. Alberta. *Help protect Alberta's beautiful elm trees.* https://www.alberta.ca/agri-news-help-protect-albertas-beautiful-elm-trees
14. Flying Squirrels. *The lord of the elms.* web.archive.org/web/20110711013315/http://www.flyingsquirrels.com/sauble_elm
15. Ibid.
16. Chang, H., Wu, J., Forde, B. & Kim, A. (2017). The Legend of Penn's Treaty with the Lenape. Haverford College Libraries. http://ds.haverford.edu/penn-treaty-elm/essays/lenape/
17. Estes, R. (2014). 'Logan's Lament'. *Native Heritage Project.* nativeheritageproject.com/2014/01/26/logans-lament/
18. Liberty Tree Society. *Herbie story.* libertytreesociety.org/pdf/HerbieStoryweb.pdf
19. Sharp, D. (2010, January 31). Do rings of Herbie the elm have age, climate data? Associated Press. web.archive.org/web/20141225022415/http://www.boston.com/news/nation/articles/2010/01/31/do_rings_of_herbie_the_elm_have_age_climate_data
20. Oklahoma City National Memorial Museum. *The Survivor Tree – Today.* https://memorialmuseum.com/experience/the-survivor-tree/the-survivor-tree-today/

Nine – Death: Elms in the Arts

1. Watt, J. (2022). *Hindsight.* Birlinn Ltd.
2. Kübler-Ross, E. (1969). *On Death and Dying.* Routledge.
3. Shakespeare, W. (2107). *A Midsummer Night's Dream.* The Arden Shakespeare.

4 Browning, R. 'Home Thoughts, from Abroad'. www.poetryfoundation.org/poems/43758/home-thoughts-from-abroad

5 Kipling, R. (1906). 'A Tree Song'. https://www.kiplingsociety.co.uk/poem/poems_treesong.htm

6 Hopkins, G. M. (1985). *Poems and Prose*. Penguin Classics.

7 Friends of the Dymock Poets. *Robert Frost*. dymockpoets.org.uk/robert-frost

8 Frost, R. (2013). *Collected Poems*. Vintage Classics.

9 Ibid.

10 King, A. & Clifford, S. (Eds.). (1989). *Trees Be Company: An Anthology of Poetry*. Green Books.

11 McKimm, M. (Ed.). (2017). *The Tree Line: Poems for Trees, Woods and People*. Worple Press.

12 Sampson, A. (Ed.). (2023). *The Book of Tree Poems*. Laurence King Publishing.

13 King, A. & Clifford, S. (Eds.). (1989). *Trees Be Company: An Anthology of Poetry*. Green Books.

14 Taggard, G. (1928). 'Dilemma of the Elm'. *Poetry 32:5*.

15 Clare, J. (1990). *Selected Poetry*. Penguin Books Ltd.

16 Duffy, C. A. (2023). *Nature*. Picador.

17 Paschen, E. (2016, January). The Tree Agreement. *Poetry*.

18 Boyd, M. J. (1923). 'Ever Vigilant: Julius V. Combs, MD' *Poetry*. www.poetryfoundation.org/poetrymagazine/poems/160514/ever-vigilant-julius-v-combs-md

19 Plath, S. (1965). *Ariel*. Faber & Faber Ltd.

20 Williams, C. K. (2006). *Collected Poems*. Farrar, Straus and Giroux.

21 King, A. & Clifford, S. (Eds.). (1989). *Trees Be Company: An Anthology of Poetry*. Green Books.

22 Dillon, G. H. (1931). The Dead Elm on the Hilltop. *Poetry 38:2*.

23 Hood, T. (2005). *The Poetical Works of Thomas Hood*. A. L. Burt Company. www.gutenberg.org/files/15652/15652-h/15652-h.htm

24 Wordsworth, W. (1994). *The Collected Poems*. Wordsworth Editions.

25 Leithauser, B. (1976). 'Dead Elms by a River'. *Poetry*.

26 Glück, L. (1995). *The First Four Books of Poems*. The Ecco Press.

27 Smyth, P. (1977). 'An Elm Leaf'. *Poetry*.

28 King, A. & Clifford, S. (Eds.). (1989). *Trees Be Company: An Anthology of Poetry*. Green Books.

29 Huxley, A. (1925). *Selected Poems*. Appleton & Company. https://www.poetryfoundation.org/poems/44299/elegy-written-in-a-country-churchyard
30 Wordsworth, W. (1994). *The Collected Poems*. Wordsworth Editions.
31 Byron, G. 'Lines Written Beneath an Elm in the Churchyard'. www.online-literature.com/byron/689/
32 Gray, T. 'Elegy Written in a Country Churchyard'. https://www.kiplingsociety.co.uk/poem/poems_treesong.htm
33 Barnes, W. 'The Elm in Home-Ground'. https://www.poetrynook.com/poem/elm-home-ground
34 Bridges, R. 'The Great Elm'. https://www.poetrynook.com/poem/great-elm
35 Arnold, M. 'Thyrsis: A Monody, to Commemorate the Author's Friend, Arthur Hugh Clough'. https://www.poetryfoundation.org/poems/43608/thyrsis-a-monody-to-commemorate-the-authors-friend-arthur-hugh-clough
36 Williams, W. C. (2004). *Selected Poems*. The Library of America.
37 King, A. & Clifford, S. (Eds.). (1989). *Trees Be Company: An Anthology of Poetry*. Green Books.
38 Gibson, W. W. (1935). *Collected Poems: 1905-1925*. MacMillan.
39 Tolstoy, L. (1982). *The Raid and Other Stories* (L. & A. Maude, Trans.). Oxford University Press.
40 Richards, D. (Ed.). (1981). *The Penguin Book of Russian Short Stories*. Penguin Books Ltd.
41 French, T. (2019). *The Wych Elm*. Penguin Books Ltd.
42 Hardy, T. (1998). *The Woodlanders*. Penguin Classics.
43 Stafford, F. (2016). *The Long, Long Life of Trees*. Yale University Press.
44 Miller, J. (2022). *The Heart of the Forest*. British Library.
45 Sater, M. (2021). Hardy's trees: ecology and the question of knowledge in *The Woodlanders*. *Nineteenth-Century Literature*, 76(1), 92–115.
46 O'Neill, E. (1995). *Three Plays*. Vintage Books.
47 Forster, E. M. (1989). *Howards End*. Vintage Books.
48 Constable, J. (ca. 1821). *Study of the Trunk of an Elm Tree*. V&A. collections.vam.ac.uk/item/O16555/study-of-the-trunk-of-oil-painting-constable-john-ra
49 Cromarty Arts Trust. *Cromarty Treescapes*. https://www.cromartyartstrust.org.uk/events/26-jul-2018-cromarty-treescapes.asp

50 Children, A. (1818). *The Waterloo Elm*. Royal Collection Trust. www.rct.uk/collection/exhibitions/waterloo-at-windsor/windsor-castle/the-waterloo-elm

51 Talbot, W. H. F. (ca. 1845). *Elm Tree in Winter, Lacock Abbey*.

52 Botanics Stories. *Elm blossom*. stories.rbge.org.uk/archives/38649

53 Puddephatt, C. *Photos of North West Scotland*. www.jacksonphotography.co.uk/elm-in-assynt

54 Clausen, G. *Elm Tree in Spring*. https://artuk.org/discover/artworks/elm-tree-in-spring-221233

55 Munch, E. (1923). *Elm Forest in Spring*. https://www.munchmuseet.no/en/edvard-munch/ekely/motifs-from-ekely/

56 Steer, P. W. (1922). *Elm Trees*. https://www.tate.org.uk/art/artworks/steer-elm-trees-n03715

57 East, A. (1908). *Autumn in Gloucestershire*. artuk.org/discover/artworks/autumn-in-gloucestershire-45979

58 Moore, H. (1934). *Hole and Lump*. https://catalogue.henry-moore.org/objects/13751/hole-and-lump

59 Hepworth, B. (1946). *Pelagos*. barbarahepworth.org.uk/sculptures/1946/pelagos/

60 Hepworth, B. (1946). *Single Form (Dryad)*. barbarahepworth.org.uk/sculptures/1946/single-form-dryad/

61 Hadzi-Vasileva, E. (2021). *Gilded Elm*. www.elpihv.co.uk/works/the-gilded-elm/

62 Lockley, M. (2018, February 25). Revealed after 75 years: The face of Bella in the Wych Elm. *Birmingham Mail*. https://www.birminghammail.co.uk/news/midlands-news/revealed-after-75-years-face-14329271

Ten – Life: Lessons from the Elms

1 Abram, D. (1997). *The Spell of the Sensuous*. Vintage.
2 Milman, O. (2022). *The Insect Crisis*. Atlantic Books.
3 Shrubsole, G. (2022). *The Lost Rainforests of Britain*. William Collins.
4 Tudge, C. (2005). *The Secret Life of Trees*. Penguin.
5 Stafford, F. (2016). *The Long, Long Life of Trees*. Yale University Press.
6 Somerville, R. (2021). *Barn Club*. Chelsea Green Publishing Company.
7 Seddon, M. & Shreeve, D. (2024). *Great British Elms*. Kew Publishing.
8 Somerville, R. (2021). *Barn Club*. Chelsea Green Publishing Company.

9 Martín, J. A., Sobrino-Plata, J., Rodríguez-Calcerrada, J., Collada, C. & Gil, L. (2019). Breeding and scientific advances in the fight against Dutch elm disease: Will they allow the use of elms in forest restoration? *New Forests*, 50, 183–215.
10 Kauri, V. (2012, May 29). Cloned Ontario tree gives hope for Canada's decimated elms. *National Post*. nationalpost.com/news/canada/cloned-ontario-tree-gives-hope-for-canadas-decimated-elms
11 Forest Research. *Elm yellows (Candidatus Phytoplasma ulmi)*. www.forestresearch.gov.uk/tools-and-resources/fthr/pest-and-disease-resources/elm-yellows-candidatus-phytoplasma-ulmi

Appendix 1 – 'Elm tree talk'

1 Fragnière, Y., Song, Y. G., Fazan, L., Manchester, S. R., Garfì, G. & Kozlowski, G. (2021). Biogeographic Overview of Ulmaceae; Diversity, Distribution, Ecological Preferences and Conservation Status. *Plants, 10*, 1111.
2 Hipp, A., Whittemore, A., Fuller, R. S., Brown, B. H., Hahn, M., Gog, L., & Weber, J. A. (2022). Phylogeny, biogeography, and classification of the elms (Ulmus). Zenodo. https://doi.org/10.5061/dryad.hmgqnk9g4

RAISING READERS
Books Build Bright Futures

Dear Reader,

We'd love your attention for one more page to tell you about the crisis in children's reading, and what we can all do.

Studies have shown that reading for fun is the **single biggest predictor of a child's future success** – more than family circumstance, parents' educational background or income. It improves academic results, mental health, wealth, communication skills and ambition.

The number of children reading for fun is in rapid decline. Young people have a lot of competition for their time, and a worryingly high number do not have a single book at home.

Our business works extensively with schools, libraries and literacy charities, but here are some ways we can all raise more readers:

- Reading to children for just 10 minutes a day makes a difference
- Don't give up if your children aren't regular readers – there will be books for them!
- Visit bookshops and libraries to get recommendations
- Encourage them to listen to audiobooks
- Support school libraries
- Give books as gifts

Thank you for reading.
www.JoinRaisingReaders.com

www.ingramcontent.com/pod-product-compliance
Lightning Source LLC
LaVergne TN
LVHW041750060526
838201LV00046B/963